BEYOND THE RANGES

BEYOND THE RANGES

Five Years in the Life

of

HAMISH MacINNES

LONDON
VICTOR GOLLANCZ LTD
1984

First published in Great Britain 1984
by Victor Gollancz Ltd,
14 Henrietta Street, London WC2E 8QJ

British Library Cataloguing in Publication Data
MacInnes, Hamish
  Beyond the ranges.
  1. MacInnes, Hamish   2. Mountaineers
  —Scotland—Biography
  I. Title
  796.5'22'0924   GV199.92.M32

ISBN 0–575–03512–9

Typeset at The Spartan Press Limited, Lymington, Hants
and printed in Great Britain by
St Edmundsbury Press, Bury St Edmunds, Suffolk.
Illustrations originated and printed by Thomas Campone, Southampton

AUTHOR'S NOTE

This book spans a period of five years and it is a record of some of the activities in which I was involved during this time.

In the chapter 'Chicken in a Basket', certain names have been changed to preserve individuals' privacy.

The passage quoted at the end of Part I, and the accompanying map, are from Vol. II of *Notes of a Botanist on the Amazon and the Andes* by Richard Spruce, edited and condensed by Alfred Russel Wallace, Macmillan, 1908. I am grateful to the Alpine Club for the use of their copy. The letter which follows this extract is printed by permission of the Royal Bank of Scotland.

The verse on page 3, from 'The Shooting of Dan McGrew' by Robert Service, is reprinted by permission of the Estate of Robert Service; and that on page 18, from 'The Love Song of J. Alfred Prufrock' taken from *Collected Poems 1909–1962* by T. S. Eliot, is reprinted by permission of Faber and Faber Ltd and Harcourt Brace Jovanovich Inc.

I should also like to take this opportunity to say thanks to my friends who shared in these escapades and who put up with me on mountain and river, and in the jungle; and I am indebted to Miss E. Whittome and Miss E. Brantley who checked the manuscript.

H. MacI.

Contents

List of Illustrations

(all photos by the author unless otherwise credited)

Following page 64

The author at Silver Camp (*photo Joe Brown*)
Llanganati: negotiating a difficult jungle section
Llanganati: forging through the arrow grass, Cerro Hermoso beyond
Joe Brown crossing the Rio Negro
The approach to Cerro Hermoso from the west (*photo Joe Brown*)
Mo Anthoine on the Rio Mulatos (Paracuma)
Expedition members at the Rio Negro

Following page 80

Miami: Joe in jail
The summit crater of Cotopaxi
Market scene near Vilcabamba: Joe in the background
Temporary hold up on the Ecuadorian highway
A breather in the Amazon jungle
Trout Camp
At the Great Landslide
Protection from the sun on the first Llanganati trip (*photo Yvon Chouinard*)
The Swamp at Yanacocha
Buachaille Etive Mor, showing Crowberry Gully
A balloon is laid out on the summit of Ben Nevis (*photo David Lewis*)
The cliffs of Ben Nevis

Betsy Brantley's double, the author, Betsy Brantley, Fred Zinne-mann, Giuseppe Rotunno (*photo © The Ladd Company. All rights reserved*)

Ben Nevis: the summit plateau showing the camp and John Poland's helicopter

Ben Nevis: Charlet skiing in gully

Ben Nevis: tents on the summit (*photo John Poland*)

MAPS

INCA GOLD

I

The Beautiful Mountain

. . . Were you ever out in the Great Alone, when the moon was
  awful clear,
And the icy mountains hemmed you in with a silence you most
  could hear;
With only the howl of a timber wolf, and you camped there in the
  cold,
A half-dead thing in a stark, dead-world, clean mad for the muck
  called gold . . .

from *The Shooting of Dan McGrew* by Robert Service

IN THE HEART of Ecuador lies a mysterious region called the
Llanganati. This is an area steeped in superstition, a land of legend and
tales of treasure. It is hostile country, as inaccessible as anywhere on
earth, where mist, dense as glass fibre, obliterates the mountains for
months on end. To the north, standing as self-appointed sentinel to
this wilderness, is Cotopaxi, the world's loftiest active volcano:
almost 20,000 feet, snow-covered and quietly smoking its pipe of
peace since its last violent outburst in 1877, when it caused total
darkness in the capital, Quito, 35 miles away. Dust from this eruption
fell on steamers in the Pacific 200 miles from the volcano!

Joe Brown, Yvon Chouinard and I decided on a vacation in this
far-flung place, where package tours haven't yet ventured. When I
first suggested this working holiday to my two friends, they were
enthusiastic. I talked to Joe about it first.

"I've been thinking about a book I read a few years ago, Joe. I was
waiting for the weather to clear on a climb near the Great Stack of
Handa, and I had this volume with me. It's about an Inca treasure, and
there's an incredible amount of loot involved — millions!"

"Oh?"

"Yes, it was stashed away in Ecuador when the Spaniards bumped
off Atahualpa, the Inca King. He was being held to ransom by the
Conquistadors; the King tried to make a deal and promised that he

would have his prison cell filled with gold if they would release him.
The Spaniards, being gold orientated like most of us, agreed, but
over the months Pizarro, the Conquistador OC, grew impatient
when deliveries weren't kept up to schedule. King Atahualpa was
garrotted and large caravans of melted-down gold ornaments, which
were *en route*, were hidden. One such consignment was stashed away
in the Llanganati."

"Sounds interesting, when do we go?"

"December."

"Fine."

"Shall I ask Yvon?"

"Good idea."

That's how our first Llanganati trip started. I wrote to Yvon, who
is even more casual than Joe about such enterprises.

"I'll meet you in Miami," he wrote, "and I'll bring a few packets of
freeze-dried food and a light stove. Take care. . . ."

Later, I was able to give Joe and Yvon the benefit of my
researches. As indicated, this treasure story started a long time ago,
on 16 November 1532.

Pizarro, with only 170 men, came down from the high passes of
the Andes and set up camp at a village on the plains of Cajamarca in
Peru. Nearby, in the hills, was King Atahualpa with 80,000 soldiers.
It was before the advent of Queensberry Rules, and General Pizarro
wasn't averse to a few cunning tricks. He sent his half-brother,
Hernando, and his henchman De Soto to the King to persuade
Atahualpa to meet with him. Next day, in good faith, the King was
carried into the village square accompanied by thousands of unarmed
warriors. Pizarro's battle cry of "Santiago" rang out and the Conquistadors fell upon the unsuspecting Incas in relentless slaughter,
capturing the King. The fall of the Inca Empire is one of the great
tales of military history.

Atahualpa bargained for his life by offering to fill his prison cell with
gold and two smaller rooms with silver as ransom. Pizarro grew
impatient and the King was dispatched to join his late sun-
worshipping subjects; then Pizarro went off to take Quito and Cuzco.

Meanwhile, the gold caravans were relentlessly heading towards
Cajamarca. The Inca General Ruminahui, who was in charge of all
consignments from the north, ordered the main caravan back to
Quito. He then marched with a vast army to meet Sebastian de
Benalcazar at Tiocajas.

The Spanish horses and their superior weapons were too much for the Incas, and Ruminahui hot-footed it back to Quito, where he razed the city and headed south with the huge treasure caravan, to Pillaro, his birthplace. He knew the area and had obviously thought of a suitable cache for the treasure.

All this is recorded by a Spaniard, Oviedo, the first chronicler of the New World. He wrote, "60,000 *cargas de ore* [loads of gold, 750 tons] were escorted out of the city by 12,000 armed guards."

Though Ruminahui succeeded in stashing away the gold in the inhospitable Llanganati, he was captured, tortured, and eventually burned at the stake in the Plaza of Quito. However, he never divulged the location of the treasure to Benalcazar.

As I mentioned to Joe, I first saw a reference to the treasure whilst browsing through a book in north-west Scotland. This was Richard Spruce's *Notes of a Botanist on the Amazon and the Andes*. Spruce, a self-taught scientist, had collected for Kew as well as other establishments. He was also asked by the Indian Government to locate the cinchona plant from which comes quinine.

Though Spruce spent fifteen years collecting in Amazonia and the Andes, he never did write up his diaries, and this was later done by Alfred Russel Wallace (co-founder with Darwin of the theory of evolution). The two volumes of this work were published after Spruce's death.

In the last chapter, entitled 'Hidden Treasure' (reproduced here at the end of this section on p. 90), he describes how he came across details of this vast fortune when he was recovering from fever in the Riobamba area of Ecuador.

Francisco Pizarro had brought the Inca Empire to its knees. In the eighteenth century, Valverde, a hard-up Spanish mercenary, married an Indian girl in Latacunga, a small dusty town just south of Quito. Her old man, who was a *cacique* in the close-by village of Pillaro, took the young Valverde into the Llanganati and showed him the whereabouts of a cave where a large amount of Inca treasure was hidden. This was his wedding present to his son-in-law.

As might be expected, Valverde came up in the world and eventually returned to Spain a wealthy man. He must have been patriotic, for on his deathbed he wrote a description of the location of the cave for the King — *El Derrotero de Valverde* — and how to get to it in the dreaded Llanganati.

The King, who was obviously not averse to a quick buck,

immediately dispatched a Royal Warrant, a *Cedula Real*, to the Corregidors of Latacunga and Ambato (both places close to the Llanganati on
the western side). This expedition of the Corregidors, like so many
which were to follow, was a failure. A friar called Longe accompanied
them and disappeared, allegedly drowned, although this is in doubt.

In those early days the Llanganati was more frequented than it is
now, with 'roads' and various mines. Its reputation as an evil place
stems mainly from the weather, swamps, the fever in those early days,
and the dense jungle to the east. As the trade winds from the east
merge with the cold mountain air, fog, cloud and often constant rain
persist.

Countless expeditions followed over the centuries. Many men went
mad, many got lost and perished. Many were incompetent, others
were highly skilled in bushcraft. Eugene Brunner, a Swiss-German,
was one of the latter. He spent 42 years looking for the treasure, and on
one of his many abortive expeditions he saw the sun only once in 127
days. Until he died in 1980, Brunner was still convinced that the
treasure, or most of it, remained in the Llanganati, and his hot
favourite area was round the highest peak in the region, Cerro
Hermoso. Though this does not tie in with the 'treasure trail' as
depicted on the old map, he did have good reasons for his choice. This
was based on an expedition in April 1887 by Barth Blake, an officer of
the Royal Navy. Blake, with a colleague called Chapman, was caught
in a snowstorm on the slopes of Cerro Hermoso, and Chapman died.
Blake managed to struggle back to Pillaro, almost dead with hunger,
exposure and fatigue. But he had with him eighteen pieces of treasure.
When recovered, he sailed from Guayaquil to Britain to organize a
fully equipped attempt to collect more treasure. Later, on his return to
Ecuador, he 'fell overboard' in the Atlantic. There was speculation at
the time that he had been murdered because he carried information
about the treasure's location.

However, Blake was a canny individual and had sent copies of his
documents to a colleague in New England; these were only discovered
many years later by his friend's grandson when he was spring-
cleaning the attic. The grandson sent the papers to Commander
George Dycott, a renowned explorer — he had gone to hunt for the
missing Colonel Fawcett in the Matto Grosso. It so happened that
Dycott was then — in the 1940s — living in Ecuador and he took an
expedition into the Llanganati. There was to be a 50/50 division of the

loot if he found it. But Commander Dycott was already an old man and he eventually gave up the search. As Brunner had been associated with him, the Swiss-German carried on the good work for many years until his death (by natural causes) in Quito.

There are numerous bizarre accounts of attempts to find the Valverde hoard. In the 1920s, for example, a husband and wife expedition was caught in a flash flood and had to scale a tree to survive. The wife died of pneumonia and her husband, who managed to get out of the Llanganati alive, went mad and spent the rest of his days in a mental institution.

I wasn't the first Scotsman to catch the Llanganati gold fever. A Captain Erskine-Loch led expeditions into the region during 1935–37. His book, *Fever, Famine and Gold*, makes exciting reading. On a later expedition, we were to follow part of a route taken by Loch into the Amazon basin, where one of his party was swept away on a dangerous river crossing — a river we had trouble with also. Other members of Loch's party went down with fever. Captain Erskine-Loch eventually returned to his jungle home at Buigraand, and with due ceremony lit two candles, placing one at either end of his table. Then he drank a bottle of whisky, took his army revolver and blew his brains out. He hadn't found the gold.

Both Joe and Yvon had assimilated most of this information before we all met up in Miami.

At first glance, the description and the old maps seemed as clear as a route card on how to find, say, Fort Knox, but there are invariably snags in a trip of this sort, and our Inca adventure was to be no exception.

Quito, built on the hearthrug of great volcanoes, is a pleasant capital city. It is high — 9,350 feet, not too hot, and the air is (as yet) unpolluted. It isn't, I discovered, a good place for budding treasure hunters to linger. One is liable to get grandiose ideas from the amount of gold that's literally lying around. For example, the huge main altar of the church, La Compania de Jesus, is partially made of gold, and the remains of the Quitonian Saint, Santa Mariana de Jesus, rest on this altar — in a gold coffin.

"Do you realize," I said to my friends as we went back out into the sun-drenched street, "that fancy piece of metalwork could finance expeditions for the rest of our lives?"

.

Our immediate destination was Ambato, a town about two hours'
bus ride south from Quito; two hours of hell where bus drivers play a
kind of dodgem Russian roulette. Glancing over to the seat in front of
me, where an Indian was reading the local newspaper and clutching a
broody hen, I read that 417 people had died on the roads in the
previous twelve months. I felt like brandishing the paper in front of
our driver — "Hey, Jimmy, have a look at this."

We stayed in the Residencia Ambato. To describe this establishment as filthy would be to do it an injustice. Both Joe and Yvon agreed
with me that prisoners would go on hunger strike if incarcerated
there. It would have made an ideal movie set for the opening scenes in
a revolutionary film. The walls were cracked, windows broken and
unsavoury, heavily moustachioed characters hung about outside our
padlocked door. Some practical joker of a plumber had managed to
arrange the toilet to flush vertically upwards, much to patrons'
discomfort, and there was only second-hand bog paper available. Like
many such 'conveniences' in Ecuador, a panoramic viewing window
with no glass was provided along one wall, which allowed one to
converse with those washing at the tap three feet away. Under my
bed, which was apparently designed for guests with acute spinal
disorders, lived an evil-looking rat.

I don't know what the Ecuadorians thought of us *gringos*. Yvon,
who had travelled the length of South America some five years
previously, is America's best-known ice climber. He can scale vertical
and overhanging ice as easily as a fit person can climb to the upper deck
of a double-decker bus. He is a quiet American of French-Canadian
extraction and was at one time a private eye. Joe, on the other hand, is
possibly the best-known rock climber in the world and has done more
new routes than most people have had eggs for breakfast. Both keen
fishermen, neither had much opportunity on this trip to indulge in
their favourite pastime, but at the Hotel Ambato Joe did put his fly-
casting to good use when he hooked his towel, which had blown from
our crumbling window on to a lower corrugated-iron roof.

We had been told in Quito that there were no modern maps of the
Llanganati, but here in Ambato we were assured that a new aerial
survey series was obtainable in the capital. I lost the toss and resigned
myself to the return journey to Quito. We knew that with a map there
was a fair chance that we would return from the Llanganati; we had
also been informed that the Indians on the western fringe of the
mountains didn't jump for joy when asked to travel into the area.

This fact became only too clear a few days later when we reached a remote village called San Jose. Dust raised by cowboys rolled down the wide, mud-baked street, the only street, where old men and children, obviously in no hurry, awaited *mañana* under ramshackle verandahs. The houses looked as if a gentle Scottish zephyr would reduce them to matchwood.

This, the highest village, was the only place to find the Indian guides who were, we were told, essential for getting into — and, more importantly, for getting out of — the Llanganati. We soon discounted that theory; not only were they reluctant to accompany us, but they demanded exorbitant payment with fringe benefits and bonuses which we certainly couldn't afford. Both Joe and I had prior experience of dealing with the Indians of Guyana, for we had climbed Roraima, Conan Doyle's 'Lost World', a few years before. There, the Amerindians were true forest gentlemen, who could move with the stealth of a bushmaster and carry enormous loads faster than we could run. These Indians of the High Paramo of Ecuador were far-removed degenerate cousins. We felt dejected.

"Hell," Yvon said, a great advocate of 'when in doubt do it yourself'. "Let's carry our own gear; we've got enough freeze-dried food and we can all get in one tent. Let's smash this closed-shop franchise and be the first guideless white men into this goddam territory."

Joe and I were still reluctant, for we had lingering memories of Amazonian rain forest and almost impenetrable walls of bamboo, prickly pine, scorpions and spiders as big as dinner plates. But we agreed; there was no alternative. After all, we had two air survey maps which covered half of the Llanganati and I had a compass — what more could good Boy Scouts wish for?

I've often thought that an ideal expedition would be comprised of Trappist monks, where there would be no arguments. Our trio was probably the next best combination. We have known each other for years and never argued, except in good-humoured debate; but we realized that we were going to need all our patience and energy.

The next day we took a hired truck up a winding snake of a dirt road, past the last *hacienda* where fighting bulls are bred, to the end of the trail. Ahead was the lake where the Incas reputedly dumped treasure so many moons ago, and across the divide to the east lay the Llanganati and, for us, the unknown.

We met a cowherd who told us that a trail went between two peaks; from there on he had no idea — only that it was easy to lose it. No, he didn't want to go with us, he never wanted to see the other side of the range — there were too many bad stories about the Llanganati.

But I, for one, wasn't to go dashing off over the hills and far away that day — I had sunstroke. Soon I was shaking all over and running a high temperature. Joe and Yvon hurriedly pitched the tent for me whilst I struggled into a sleeping-bag, with the temperature at 105 degrees in the shade. Next morning I was still cocooned and feeling as weak as a Bangladesh kitten. It wasn't until the following day that I managed to pick up my heavy rucksack and stagger along behind my two friends.

We were very close to the equator and it was incredibly hot. The sun burned the skin on the back of our hands and even the cacti appeared to perspire. The weather was unusually good, for the Llanganati is normally enveloped in cloud — this, we had been told by the Indians, was the best weather for years. Despite our resembling pork crackling, we could count ourselves lucky.

When we resolved to do our Baden-Powell stunt and go it alone into this great beyond, we also had to decide on other aspects of the treasure hunt. We knew from our past experience of jungle country that three of us would be unable to spend weeks trail-cutting through bush in search of the Valverde gold mine. It just wasn't possible to carry enough food, even of the dehydrated variety. Fuel was another problem. There was no combustible wood in the Llanganati, or so we had been led to believe, but that was another theory which we later sent up in smoke. We came to the conclusion that if we headed for the highest peak in this wilderness, Cerro Hermoso, we could then perhaps ease off northwards to reach the general area of the mine as shown on the Valverde map. Studying the new maps, we reckoned to have found a route to this peak along high ridges which probably had less undergrowth on them.

"After all, that Englishman who was bumped off— Barth Blake — found a couple of gold breastplates or some such goodies on Cerro Hermoso," I encouraged my two friends.

"Is that right?" Yvon responded.

"I wonder if there really is any treasure left in this wilderness?" Joe mused.

When we reached the divide on the fringe of the Llanganati, we saw ahead of us rough, uncompromising country — not the place to take one's mother-in-law unless one wanted rid of her. The greenery was punctuated by lakes and mountains. Stunted sterile-looking peaks merged in the far distance with the forest of the upper Amazon. We plunged down into a green hell, for hell it was: the arrow grass — well named, it is used to make arrows — was as dense as a wire brush. In places it was 20 feet high, and there was a danger on the steep slopes of spiking an eye on the hard-tipped leaves.

I was weak in the morning when we started, and soon felt exhausted. We crossed a stream where, with more dedication than enthusiasm, I prospected with my gold pan.

"Bugger-all here, lads," I announced, replacing the pan (really a plastic washing-up bowl) in my pack. "But I'm glad of the rest."

This was a barren area; there weren't even any insects. Indeed, during our stay in the Llanganati we saw only one bee. It seemed overjoyed to see another form of life, for it buzzed around us for ages as if we were old friends. On this *terra firma*, there are pumas, tapir, and the odd large deer. Aloft, the condor spirals in lazy aloofness.

It was about 3.30 that I collapsed in a clearing the size of my back porch and said that I could go no further, so we made camp.

"It's going to be chilly-billy tonight," I muttered, my teeth beginning to chatter. "Can I sleep in the middle?"

"Sure," Yvon drawled in his slow Californian accent. "Be good and don't snore."

"I don't snore and I don't fart — very often. Not like Brown here. He's well named."

"Shut up, you Scottish git," he returned. "Go to your pit before you snuff it."

Within the hour a bitter frost descended, enveloping our world in deep hoar. A full moon soared over Cerro Hermoso.

The next day I felt stronger, and we made better time. Hitting an old animal trail on a ridge, which went in the desired direction, we walked all day in broiling heat, with handkerchiefs Ku-Klux-Klan-like, covering our faces for protection from radiation. I prospected on the way, but didn't find a glimmer of gold though there were piles of iron pyrites: fool's gold. We then managed to make a fire, otherwise we would not have eaten as we had only a limited supply of small alcohol-survival-cookers. I shudder to think what might have happened if the usual rains had been in evidence.

A knife-edged ridge led to Cerro Hermoso — like a cathedral roof. It was back-breaking toil sweating along it but each step took us closer to this elegant spire of a mountain. We stopped after a steep pitch. To our left was an uninterrupted view across the peaks of the Llanganati,

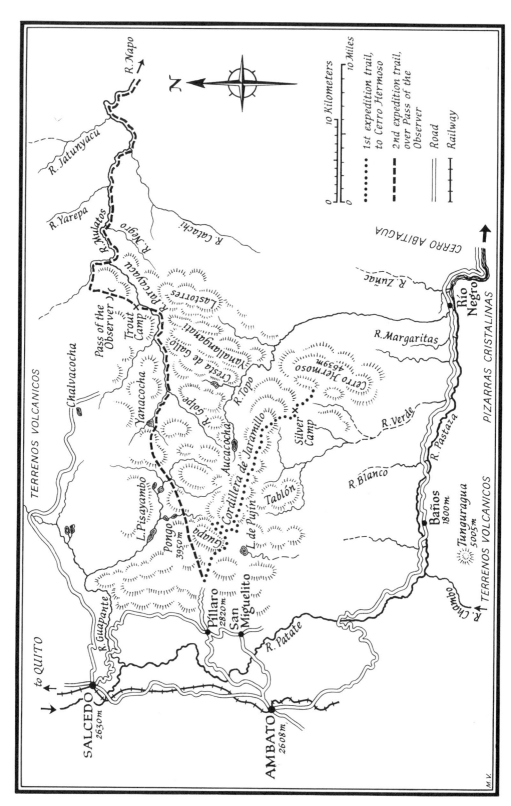

The Llanganati of Ecuador

with the great carpet of Amazonia in the hazy-blue distance. Below us was a small lake, then another larger one in the middle distance. Yvon offered Joe and me a sweet.

"Thanks," I said, feeling weak and tired.

"Not much sign of wild life," Joe observed, popping the caramel into his mouth. "Yet there are plenty of trails."

"I read somewhere," I ventured, "that the rare hairy tapir makes trails like those."

"I suppose that's better than hairy Scotsmen," he replied with a laugh.

I ignored this remark and continued with my useless information. "Also, there's the spectacled bear, another rare chap, and Joe, before you get it in, it doesn't wear bifocals."

"Anything else?" asked Yvon quietly.

"That's about it, as far as I've read — oh, there were wolves here before the last war, but they may have gone now. They mainly lived on rabbits. Apparently, they surround the rabbits whilst they are nibbling their breakfasts down by the swamps, then, on a given command from the pack leader, they make a concerted attack. The poor bunnies are snaffled up before you can say snap, crackle, pop."

For several days we remained at an altitude of 14,000 feet, but nowhere could we find a route which would take us north towards the probable area of the gold mine without the greatest difficulty. Eventually, however, we drew close to the slopes of Cerro Hermoso, a savage, naked mountain elbowing above its neighbours. There is silver on Cerro Hermoso and possibly gold also, so I had more than a casual interest in this mountain. We hoped besides that the valley leading from it towards the Rio Napo, a tributary of the Amazon, might provide us with our golden short-cut.

It was now Joe's turn to be ill. He suffered from dysentery, a malady which was to remain with him for most of the trip. We set up a camp about four hours' march from the mountain, at a point where there was a stagnant pool. This provided a good base from which to explore and prospect.

The next day, I was to have an alarming experience when out prospecting alone. The trails which we used appeared to have belonged to a race of mini rabbits which had all passed away — no doubt due to the wolves. Their paths were certainly minute and very difficult to follow. I invariably used the Hansel and Gretel technique, leaving sweet papers at difficult points on the trail. But Yvon, like most Americans, is so litter-conscious that he can't bear to see such tokens of civilization and insecurity lying about, so he surreptitiously pocketed the offending wrappers.

That day we started out at sparrow fart and, at the base of Cerro Hermoso, found an old camp of thatched huts.

"I wonder if this was the camp used by Brunner, that Swiss-German guy?" I asked.

"Quite a place," Yvon observed.

"If so," I continued, "he came in from the south with his expeditions. The town of Baños lies down there." I pointed down over the bush-clad slopes to a deep valley. The ghost camp had an eerie feeling about it. There were ancient maps and bundles of wet dynamite. The camp squatted at the brink of a deep lake which we called Silver Lake, and the cliffs of Cerro Hermoso rose above in a great cirque. It was a geologist's paradise. There were enormous bands of gneiss and iron and the summit of the peak was of a black shale.

Joe and Yvon climbed the peak whilst I made a thorough survey of the streams in the area. I found traces — sparse — of gold and there was also copper in possibly workable quantities. Then I tried to cut down the valley leading north towards the Valverde mine without success; it would be a hopeless task without the back-up of a strong, well-equipped party.

I lost my way returning to camp. I could have sworn that I had placed a yellow sweet paper to mark an obscure turning-point on the path, but I couldn't find it and retraced my steps five times, but always arrived at a clearing where a menagerie of pumas had obviously had a dance, for there were pug marks everywhere. In the end I had to resort to my compass and, taking a backbearing on Cerro Hermoso, cut through the arrow grass to arrive, much to my relief, back at camp. Joe, who can follow the faintest suspicion of a jungle trail as competently as a Javaro Indian, was amazed at my predicament.

Round the campfire that night we weighed up the pros and cons of continuing from this part of the Llanganati to the Inca mine, as shown on the old map. It was obviously hopeless; there was no way we were going to reach the general area of the mine or cave. Neither had we any chance of going in on the route of the original description; that, too, would require a large and well-equipped team. Such an expedition would need Indians from the Amazon basin to ensure success and to lure them from their rain forest haunts just wasn't on.

"What was the climb like?" I asked finally. We were eating a freeze-dried meal and long shadows played in the small clearing. I could see the Southern Cross and this reminded me of my days in New Zealand.

"Bloody gruelling," Joe said, licking his spoon. "Especially as I was still weak from the shits."

"Sure was quite a hike, Hamish and, do you know," Yvon looked at me quizzically, "we came across some very deep holes; they went straight down into the rock, didn't they, Joe?"

"Yes, they did, for hundreds of feet," he agreed.

"I wonder if they could have been vents?" I suggested. "I met a geologist who told me that, despite early beliefs, this area is probably volcanic. He had heard of those holes, too. And that's another thing that's bugging me," I confessed. "I was thinking about it as I was panning and scouting about today. There seems to be some confusion about this whole treasure business. Is it a mine we're looking for, or a left-luggage cave with Inca loot?"

"You tell us," Joe invited. "We believe everything you say."

I ignored this remark and continued. "Well, according to that old Spanish scribe, Oviede, General Ruminahui took his sixty thousand loads of gold and twelve thousand guards and stuffed the gold in a cave. I'm thinking that there couldn't have been many moles in those days — they probably didn't keep their heads if they shot off their mouths. Anyhow, what I'm trying to say is this: there may be two lots of golden goodies in these here hills. The cave, which Blake and Brunner thought was here" — I jerked my thumb in the direction of Cerro Hermoso — "and a separate mine, from which the Incas got a lot of gold. It makes sense, for the Valverde guide to do-it-yourself riches mentions mining gold and even describes a primitive smelter."

"I guess we'll never be in a position to verify your theories, Hamish. We've still got to get the hell out of this damned place!"

"You're right, Yvon," I responded. "We're out of grub and this weather can't hold. We've had the luck of the righteous, which we're not."

"Oh, well," Joe interjected. "It's back along that rollercoaster of a ridge again to the fleshpots of Ambato . . ."

For the next few days we retraced our steps along those narrow ridges. It was a different viewpoint now. On the second day we dropped down into a basin of arrow grass and, following the trails of our friends the tapirs, we got lost.

On wide swampy ground we had a meeting of the expedition committee. We all took opposing views, each adamant that his direction was the correct one, and each direction about 90 degrees apart.

"I'm going this way," Joe decided, and impetuously set off.

Yvon and I tagged along, perhaps apprehensive of spending the rest of our days contemplating confinement in a patch of arrow grass, which resembles a closely-barred 360-degree window.

Presently, when it was obvious even to Joe that we were heading for distant Venezuela, I pulled out my compass.

"Look here, Brown, you're taking us north-east; we want to go due west."

He had to agree when he looked at the compass and, after what seemed an endless day, we crested a line of hills and from there saw the high paramo which eventually takes one to Pillaro.

It was three grime-covered men who arrived back in Ambato. Our lips were cracked and bleeding from the sun and our clothes ripped and black from campfire smoke.

"No," we told enquiring Indians, "we didn't find the treasure, but we think it's there." They smiled knowingly; they, too, know it's there. But I'm sure they thought that it wouldn't be found by filthy gringos.

However, we found temporary (better) lodgings in Ambato for that night, and, sitting on chairs facing each other, with a basin on the floor between us, we communally ate the most delicious pineapples in the world. Juice ran down our faces into the basin. It was bliss.

"Well," Yvon remarked, replete at last, "if we can't find the goddam Inca treasure, at least we can climb mountains. From past experience," he continued, "you have to do everything as soon as possible in Ecuador, before disease and lurgy catch up with you. Let's bag one of the big mountains before we're carted off home on a pill-laden stretcher."

"Cotopaxi," I said, knowingly, "is the highest active volcano in the world. How would that do for a day's outing?"

"Okay," Joe agreed, "let's do it."

As I said, we didn't disagree fundamentally and our tastes were similar (we all liked pineapples).

But to climb it we had to return to Quito. We realized this after waiting a whole day for a lift at a dusty roadend, where a track led in 30 miles to a remote hut on the mountain at 15,500 feet — about one vehicle a week used this road and one's hitch-hiking thumb doesn't get overworked! This exercise illustrates the perennial optimism of the Scots — it was my idea. We hired a jeep in the capital and drove past the roadend of our patient vigil to another unusual occurrence in Ecuador. If the Inca gold god didn't favour us, the weather god did, and we tumbled out of the vehicle sucking as much air as possible into oxygen-starved lungs. The hut, which seemed only an orange-throw away, took 45 minutes to reach, and here an Indian custodian made us herb tea which tasted like nectar.

Yvon had recklessly brought a rucksack full of exotic fruit, a virtual cornucopia of vitamin C. Joe, not being a fruit bat like Yvon and myself, found that it encouraged traumatic bowel activity and caused profound distress to his two friends. That night in the hut his malady was audibly and olfactorily verified by Yvon and myself.

Cotopaxi has been on its best behaviour for the last 100 years or so, and we fervently hoped that it wasn't going to change its mind in the following twelve hours. At 3 a.m., with headlamps probing the night, we crept up its glaciated flanks, turning deep sinister crevasses and ascending its icy walls. Conditions were perfect, but our lungs, not yet acclimatized, felt the strain. It was cold, and by the time we approached the summit, the wind cut through us like the arrow grass of the Llanganati. At one stage, Joe thought he had frostbite.

It was a strange experience, arriving on top of that big mountain before dawn — Cotopaxi is almost 20,000 feet. Far over the Amazon there was a greenish diffused light and then, after all the effort, we were on the rim of the great crater, 2,300 feet in diameter north to south, with a wisp of ominous smoke curling up from the edge. Dawn was marching across the rain forest, but we couldn't linger; to do so would have meant being frozen. We headed down; down to Quito where the jeep was handed back exactly 24 hours after collecting it. A fast but memorable trip. Ecuador is a land of dust and fruit. For 10p you can buy the most succulent pineapples. Yvon and I ate our fill and thought of our next ploy. Joe looked on in envy but prudently declined to join us in our snack.

II

Peter Pan Land

I grow old . . . I grow old . . .
I shall wear the bottoms of my trousers rolled.
                                                      T. S. Eliot

AFTER OUR ABERDONIAN excursion to Cotopaxi, I suggested to my
two pals that we should take things easy for a bit.

"How about a few days down in the south of Ecuador, in
Vilcabamba?" I proposed. "It's one of those odd places where you get
Methuselahs hopping about like breadsnappers."

"I was thinking of heading back home," Yvon objected. "I've a big
business deal coming up. . . ."

"The trouble with you, Chouinard," I muttered, "is that you're
wheeling-dealing too much these days. Easy does it . . . relax . . .
take up contemplation, or even fishing." Yvon, like me, spends a lot
of free time studying imaginary horizons and he rivals Joe for an
obsession with 'tight lines'.

"Right, I'll go," he agreed resignedly. "But let's fly; I'm not sold on
bus travel here and being zooed by all those Indians."

"It's a deal. Okay with you, Joe?"

"Sure, we may even get a few tips for retirement."

"Just as well you're coming, Yvon," I remarked. "Without your
Spanish we'd be as out of place as a couple of Mormons at a cocaine
party. I'll go and get the tickets. I think we go via a rain-drenched
dump called Loja."

As I trotted over Quito's pavements to the national airline office, I
mused on what I knew of the old people of this world; Americans
would call them 'super-old', I thought.

Back in the 1950s, when I had visited the remote state of Hunza in
north-west Pakistan, I was impressed when His Highness the Mir,
with whom I was staying, pointed out several old men still toiling in
the fields.

"Two of these men," he told me, "are over a hundred years old."

There are at least three places on earth with an unusual concentration of oldies: besides Hunza, there is Abkasia in the Caucasus mountains of Russia and Vilcabamba in southern Ecuador.

I should probably count myself lucky to have visited all three places, but neither Hunza nor Abkasia has documentary evidence for the actual ages of their respective greybeards. Vilcabamba has. There is a birth certificate there, reputedly genuine, for Jose Maria Torres, aged 150 years. No such papers exist in either Hunza or the Caucasus, though Shirali Lusimov, who died in 1973 in Abkasia, was reputed to have been 168 years of age. Using a new age-dating technique, American scientists have been able to confirm the age of a Russian nonagenarian woman from one of her teeth. This method depends upon chemical changes (racemization) that take place in the amino acids of the once living protein when an animal dies. Originally, the system was utilized to date the relative ages of remains hundreds of thousands of years old. It was discovered that similar processes take place in some tissues during an animal's life — for example, with teeth and eye lenses, formed in early life and not replaced by new protein. Tests on tooth samples show that the racemization technique can pinpoint ages to within ten per cent. Perhaps in the future this method of dating 'old people' will be employed to determine the age of many of the centenarians who have no birth certificate.

These three places in the world have one thing in common: they're darned difficult to get to! I was intrigued to find out what made these old people tick. They certainly tick longer than most and the reason for this is partly their robust hearts. There is virtually no heart disease, or cancer for that matter, either in Hunza or in Vilcabamba.

To get on a plane from Quito to Loja at a weekend was virtually impossible. It was certainly impossible for us, so we had to take the only other means of transport: bus, much to Yvon's displeasure. Already on this expedition I had spent sixteen hours on Ecuadorian buses — my journey to Quito to get the Llanganati maps — and the thought of a further 27 hours on these fate-defying projectiles made me feel quite ill.

There was no alternative, however, so we made our way to Quito's bus terminal, then temporarily located in a great dustbowl on the fringe of the city. The Ecuadorians are great people for painting their public transport vehicles in gay colours. Even the national airline's jumbo jets resemble psychedelic condors.

Amidst the dust and noise, drivers and conductors extol the virtues of their particular bus at the tops of their voices. Above this hullabaloo eager Indian women, not exactly exemplary models of personal hygiene, sell cooked guinea-pig, ice cream, boiled eggs, fresh eggs, fruit, chewing gum and, last but not least, anti-travel-sickness pills. The instructions on these (in English) stated that they were excellent for disorientation. One of the best-selling pharmaceuticals was a brand of tranquillizer: a necessary prophylactic for Ecuadorian bus travel.

This is a country of free-enterprise transport and in some ways it must be comparable to the great days of steam in Britain, when rivalry was rife between various railway companies.

The journey started innocently enough. The bus was crammed with various forms of animal life — hens and geese clucked and chuckled and appeared to enjoy the constant jostling. The aisle was overflowing with Indians sporting trilbies and the ubiquitous welly boot, which is not, I observed, the monopoly of the Scottish West Highlander. It is illegal to have standing passengers on these buses and frequent stops at police posts caused a flurry as the conductor (who only dons his official hat on such occasions) shouted to all standing passengers to lie down in the aisle. The women must have found this particularly difficult because a large percentage of them had babies lashed on to their backs with shawls. But, if the laughter was anything to go by, it was apparently considered great fun. Invariably this ploy fooled the police who waved the bus on, after a perfunctory glance at the sheaf of papers listing the names of the passengers.

For the first hour or so the journey south was no worse than an ordeal with a drunken driver in rush-hour traffic. Not that the driver was drunk — but he may well have been chewing coca leaves. It was only when the opposition — in the form of rival buses — made their presence felt with deafening blasts on their air horns as they shouldered past, that our driver showed symptoms of consternation. The cherub face which he had presented as he lured us aboard was now transformed by furious grimaces as he endeavoured to squeeze more speed out of his charge. Sometimes he would glance furtively above his head to where a Madonna was hanging, illuminated by a red bulb.

As we snaked our way high into the Andes, the rain, the first we had seen in Ecuador, lashed down off the mountains. Visibility was reduced to 50 feet but there was no reduction in our break-neck pace. It was on this journey that we discovered all about Ecuadorian passenger insurance. Every so often the conductor would make a solemn announcement and take round his elegant hat. Into this passengers dropped a few coins. At the summit of a pass, the conductor, having judged to a millimetre our approach to a shrine of St Christopher (the patron saint of travellers), leapt off the bus, deposited the donation, lit a candle and sprinted back to jump aboard as we accelerated again. Ecuadorian bus conductors don't have to jog to keep fit. When our bus overtook another on blind corners we courted heart attacks. Many who take this traumatic means of transport to Vilcabamba for the relief of heart complaints must, inevitably, suffer further cardiac damage. But the passengers displayed great patronage for their driver and cheered like football fans when he vibrated past a rival bus. Hours later the driver's bloodshot eyes were glazed as he stared as if mesmerized into his large rearview mirror more often than on the road ahead, whilst constantly spurred on by the ecstatic passengers (excluding three). At the town of Cuenca, a fascinating outpost of the Inca Empire, the bus skidded to a halt and we unfolded like geriatric deckchairs, looking for somewhere to lay our weary and grateful heads. It was 10.10 p.m. We had left Quito that morning at 7 a.m.

Next day we reached Loja. Sufficient to say of this part of the journey that it lasted a mere twelve hours, but seemed to take five years off our lives. There were a couple of minor incidents. A huge truck slewed off the road and effectively blocked it, and a bus which had squeezed past ours with a cheer from its passengers on a narrow part of the dirt road (it would have been tight enough for two minis) had a head-on collision at the next bend. Fortunately it was not serious, but we had heard of similar ones where over 60 people had perished.

We counted ourselves lucky when we arrived at Loja and thanked St Christopher that we had managed to book seats on a plane for the return journey to Quito. One such bus ride was enough. Since our visit, though, three planes on internal flights in this part of Ecuador have crashed and disappeared.

Loja is a wet and smelly town. Its one 'scenic attraction' is a diminutive river which loiters down a main street — no doubt it is popular with the mosquitoes. We didn't stay long in Loja, just long enough to sample the aforementioned mosquitoes and to hire a four-wheel-drive Toyota. Looking at the fuel gauge I suggested to Yvon that we'd better get some gas.

"Hell, no," he replied. "There's plenty in the tank." Both Joe and I were not convinced by this casual statement, but didn't like to insist on a further expenditure of 30p to fill up. Petrol is very cheap in Ecuador.

Vilcabamba is a one-horse town or, more precisely, we saw two horses and a donkey. There was no aura of well-being which we had been led to believe existed — it just looked sleepy, like a tired dog. As a matter of fact we thought that the Llanganati of evil repute had more zest. Here it was hot and dusty, with flies which seemed to be on double time. The town was dominated by the church, built by a Vilcabamban who had made good. There was a corner café where, under a poster of a nubile nude, we were served solid bean soup which looked as if it had been made in a batching plant. The waitress, a sloe-eyed maiden, displayed as much alacrity as a zombie; she didn't seem a good advert for a place boasting of centenarian Lotharios.

"This goo should put lead in our pencils," I muttered, quickly tucking into the soup, lest it set.

"Dinna bend yer spoon, Jimmy." Joe wagged a finger at me.

Reflecting on the energy-promoting properties of bean soup, I know that there is no lack of virility amongst Vilcambambans. Indeed, many of the over-eighties openly boast of how many women they can sleep with in one night, so there must really be something about the place — or the beans! Eyeing the small village square, we came to the conclusion that the centenarians must be recuperating after their nocturnal calisthenics. The oldest person in sight was the village bobby who looked like a limp noodle. He was complete with a soporific horse and rusty six-shooters and didn't appear overworked.

The name Vilcabamba means 'sacred valley' and it is certainly world renowned as the place for longevity. Cardiologists have beaten a trail here, but have themselves fallen victims to the lack of hygiene. Over the years the old people of the valley have built up a resistance to germs, but despite investigations, scientists haven't yet pinpointed the reason for the great life-span in Vilcabamba and the adjoining villages.

At 4,500 feet, the village of Vilcabamba is close to the height of Hunza and Abkasia. I know from personal experience that all three places enjoy a very dry environment and that there is an abundance of fruit. Also the people continue working even at a great age. For example, close to Vilcabamba, Victor Maza, 120 years old, was tilling his own little patch of land until quite recently. These natives of the sacred valley don't eat much meat; they're too poor. Beans, corn, some cheese and potatoes are their staple foods. Many smoke and indeed grow tobacco in their gardens; they drink both coffee and rum, the latter made from sugar cane grown in quantity at the expense of fruit trees which were decimated to make way for this crop. But it is not the excess of sugar which makes the Vilcabambans' teeth so bad; even young children are affected and dental decay was prevalent long before sugar became popular in the district. There is a strong family bond amongst these villagers, as well there might be — they've had plenty of time to work on it. For example, Miguel Carpio, 123 years old, had twelve sons, 98 grandsons, 78 great-grandchildren and 72 great-great-grandchildren.

I can recall getting quite merry myself in Hunza where the Mir plied me with Hunza Pani. The Russian centenarians are also partial to a drop of wine, so some may say the elixir of life may be bottled, but I don't think so. Longevity is more likely from the hard work and relatively fat-free diet and a congenial climate.

Vilcabamba is certainly a far cry from the United States where a million people a year die from coronary diseases. In Vilcabamba men appear to live to a greater age than women, though one couple — the Carrions — grew old together, she being 140 and her husband 145. Many of the men have been married two or three times at 120 years of age and over. The women of the valley are usually underprivileged and overworked, being treated as sexual playthings at feasts and weekends. The old men often take a local fertility drug called gayuna. They insist that it improves their drive immeasurably. The women make no such claims for its potency but many women of sixty bear children. The centenarians of Vilcabamba haven't heard of health freaks. As well as drinking and smoking, many take drugs such as wilco from the wilco tree which 'sends' them for hours on end.

One of the reasons why we were interested in this part of the world is that nearby, in the valley of Palmyra, there are stories concerning another Inca treasure. On a hill called Sercumluma just south-east of Vilcabamba there are broken statues and traces of Inca and pre-Inca stonework. The last of the Inca kings, Atahualpa, whom I mentioned in connection with the Valverde treasure of the Llanganati, spent some time in the valley of Palmyra beyond the ruined town. It was here that he started his initial bargaining with Pizarro and his Conquistadors. As every schoolboy knows, he was carted off to Peru. The treasure which was then currently *en route* (as was the Valverde treasure) to pay his ransom demand was either hidden or disposed of by the transport division of the time. It was said that there were a thousand mules carrying this part of the ransom from Palmyra under the command of the King's captain, Quinara. To this day the river which dances down the peaceful Palmyra valley is known by his name. Quinara buried the treasure in Palmyra when he heard of the King's death. He stowed it in seven shafts; included in this consignment was the King's golden throne, his private transport. Of course the canny captain marked the spot where the gold was buried and the principal 'pointer' to this hoard was an eight-foot-high statue. Reputedly, it is still there but we couldn't find it.

In those feverish days the Jesuits had a monopoly on quinine from the cinchona bark, which was originally found in the region of Malakados close by. When an Indian came to them one day suffering from malaria, it transpired that he was a member of the party that had buried the treasure for Captain Quinara. The priests, always on the lookout for loose change, managed to get the Indian to draw a map with the burial shafts marked on it. In their get-rich-quick bid they apparently disturbed certain critical marker stones which were keys for locating the shafts, so they went away empty-handed. One priest, however, remained, and after years of fruitless search went mad, but no doubt he became a proficient hole-digger.

About a couple of hundred years ago the saga took on a more tangible slant. Sanchez Orinjana, the equivalent of a Scottish crofter, was inspecting flood damage on his patch when he discovered a pile of gold ingots, presumably Inca treasure, also some priceless Inca gold ornaments, for the Incas were master goldsmiths though, regrettably, little of their art remains today. But Sanchez, not to be influenced by arty-crafty work, melted the lot down, no doubt to prevent too many questions being asked. It took 120 mules to carry his load to Quito. The humble crofter became the Marquis of Solande, Solande being part of Palmyra.

We waddled back to the Toyota as if under ballast.

"That's what I call a filling meal," I remarked, climbing into the back of the vehicle — Yvon was driving.

"Yeah, pretty solid stuff. Just like porridge," Yvon added.

"Hey, look at that petrol gauge," Joe pointed. "It's showing empty."

"It won't run on bean soup, that's for sure," I interjected sorrowfully. "Where the hell will we get petrol now?"

"I'll ask." Yvon swung out of the driver's seat and addressed himself to a young lad who was obviously returning from working in the fields. He carried a mattock and his boots were red with clay.

"There's a small shop down this side-street," Yvon told us when he got back in. "I think it's a sort of general store."

Eventually we found a ramshackle hut which looked as if it had evaded a demolition order about 100 years previously. However, Yvon went in and presently came out with two large five-litre bottles of petrol.

"They probably drink the bloody stuff, Joe," I suggested.

"Ten litres of petrol won't get us very far," Joe pointed out caustically.

"Oh, we'll find a gas station somewhere," Yvon replied optimistically. I wasn't convinced.

We set off in a cloud of dust and went over the hill to Palmyra. Our passion for gold had been somewhat quenched after the Llanganati episode, but it wasn't so hard to reach this valley. We had the Toyota, and the greatest obstacles were two rivers which we had to ford. The holes made by past treasure diggers were still there, but we couldn't see any sign of the statue, which some thought would be obligingly pointing to the burial shafts. Nor did we get round to using a sophisticated metal detector, which we had with us, for the natives here appeared sullen and hostile. We even ran out of petrol on a remote jungle track!

"This just isn't our scene," I said to Joe, who was shaking his head in disbelief at our predicament. No treasure, no petrol and, a short time before, we had been told by an Indian that one of the centenarians, Gabriel Brazo, who was 127 and whom we had been tracking down in Palmyra, had died recently.

Whilst I guarded the Toyota, my two *gringo* friends set off along the steaming trail in search of petrol. A short while later an old man came along the path towards me carrying a great bundle of firewood. I couldn't guess how old he was but he was obviously a great age, yet he had the sprightly walk of a youth, like other old people we had seen. He cheerfully passed the time of day and I looked at our useless vehicle and thought of the absurdity of progress and the complications which we bring into our lives. The old people of Vilcabamba will still be going about their business in their timeless way long after the rest of the world has used up its reserves of oil and its tolerance to its neighbours.

III

To the Woods

"No, no, I'm only thirteen."
"This is no time to be superstitious."
"I'll tell the Vicar!"
"I am the Vicar . . ."

JOE BROWN TELEPHONED me one day from his retreat in the damp valley of Llanberis in North Wales and suggested that we should return to the Llanganati of Ecuador. I was not bubbling over with enthusiasm at this offer for it was only eight months since Yvon Chouinard, Joe and I had been there and recollections of a sizzling sun which puts a British Steel furnace in the shade must have been more than skin deep.

"It's a trifle warm," I argued. "How about Antarctica instead . . . ? Yvon is keen on tracing the steps of Shackleton in South Georgia and there are some super unclimbed peaks," I added as extra bait.

"Martin Boysen and Mo Anthoine are both keen on Ecuador. If we went further east, who knows what we might find? Remember that great expanse of forest?"

"Creepy-crawly land to be sure," I retorted with as much enthusiasm as arranging for an appointment with the dentist. "But leave it with me. I'll see what I can dig up in the form of a sponsor."

Well, that is how the 1980 Llanganati-Rio Napo Expedition came into being. There was no shortage of sponsors and the BBC wanted to send a film crew with us.

Yvon Chouinard had in the meantime opened a new shop in Los Angeles and felt he couldn't make this particular expedition. Since our last treasure trip we had arranged to attempt to climb a remote rock spire in Colorado. This was to be combined with an exciting raft journey down part of the Green River with a folded balloon aboard, which was to have served as a static camera platform for filming the first ascent of Yvon's pinnacle. He had first seen this intriguing feature during a trans-American flight and had got a fix on it from the pilot.

However, subsequent surveys failed to locate it so the project was called off.

I didn't want to go back to Ecuador and have all our eggs, or nuggets, in the one basket. It's all very well hunting for treasure, but it does have a high failure-rate. Especially the Valverde hoard. Therefore I proposed that we combine the trip. Part treasure hunt and part journey into the unknown. We would try to force a route from just north of the peak Cerro Hermoso to the Orient. Though a distance of a mere 80 miles, it was probably the most difficult 80 miles on earth.

There was no shortage of recruits for the expedition; it seems amazing that there are so many foolhardy people willing to expose themselves to the privations of hostile places, but two further friends met with our joint approval: Dr Joe Reinhard, an anthropologist, was known to both Joe Brown and myself, having been with us on the yeti-cum-Lost Valley hunt in the Himalayas a couple of years before. Joe is a tall, spare, easy-going American and a specialist in Shamanism, with a curriculum vitae which reads like a *magnum opus*.

Our man on the spot was to be Brian Warmington, an Englishman and a climber, now domiciled in Ecuador in the frontier-like town of Baños which nestles as if in perpetual siesta on the southerly doorstep of the Llanganati. Brian is an hotelier and as a fluent Spanish speaker was to prove to be, if not the right arm, the tongue of the expedition.

Making a film in South America is like trundling a great snowball downhill. It gathers not only mass and momentum, but tends to rock the tranquillity of even the most friendly bureaucrats. The Ecuadorian Government was no exception. Apparently, an American TV company had made a film documentary a short time previously which had been both offensive and inaccurate, with the result that our BBC boat got rocked in its repercussions. What we thought was going to be a free-and-easy holiday, like the previous year, was now being strangled by a boa constrictor of red tape. We were told that it would take six months to get filming clearance. It was now November and I was due to leave for Quito in three weeks' time. Also, owing to an inter-union dispute, our BBC Scotland film crew were to be replaced by two unknowns from the anonymous labyrinth of the BBC London studios. We were not happy about this and were relieved when Mike Begg, the executive producer, decided to pull out. Though we didn't realize it at the time, this was the salvation of the trip, for no way could we have got a film crew down those Amazonian rivers unscathed. It would have transformed a hard but enjoyable adventure into a cacophony of squabbling porters, if indeed we could ever have persuaded them to come with us in the first place. There was also the possibility of serious injury in an area where rescue was as likely as tripping over the gold ingots of the Valverde treasure.

Before we left, I borrowed a book in Spanish on the Llanganati. I was excited by the mention of a treasure reputedly found by an Ecuadorian, Puga Pastor, a Corregidor, or magistrate, of Latacunda, a small town to the north-west of the Llanganati. The treasure was shipped from Lambayeque in Peru in 1803 on the order of the magistrate dated 1794. The fortune was then valued at £460 million and was sent on the ship *El Pensamiento* under the captaincy of John Doigg in a cargo of various chests containing gold bars, plate and gold dust. The deposit was entrusted to Sir Francisco Mollison in accordance with the instructions of the magistrate. Puga Pastor died in Lima in 1805.

It appeared to me when reading this that we had been beaten to any bucket-and-spade excavation by this wily magistrate, but we none the less consoled ourselves with the hope that a few grains of gold dust might remain.

All this came to light in 1965 from a document found in Lima and gave descendants of Puga Pastor a possible claim on the fortune, or so they thought. As I was due to head off to Quito ahead of the others, I decided to leave this aspect of the enquiry until I returned; after all, I knew that it was quite impossible to get anything done over the Christmas period.

However, the lucky magistrate of Latacunda had not been the only treasure seeker: many stalwarts had braved the mists and legends of the Llanganati in the past, for it is as much steeped in moisture as in superstition. There are few places left on earth which are safe from the ubiquitous tourists who, like penetrating oil, seep into uncharted territories in the wake of the pundits. This is one of the reasons why I find the remote parts of South America more attractive than the now overrun Himalayas.

The Llanganati is one place which has retained its identity, and the country to the east even more so. At about one and a half degrees south of the equator you would expect a sauna-like climate; indeed it is sometimes like that during the day, but when dusk falls, abruptly, as it does at this latitude, frost can bleach the wilderness. Admittedly, the Llanganati lies at about 14,000 feet, give or take a few valleys or summits, and on its easterly back garden the arrow grass gives way to a wide green ocean of rain forest. From here water is recruited for the Amazon on its 4,000-mile trudge to the Atlantic.

Few people have ventured east of the Llanganati and with good reason; trails end abruptly and rivers accelerate as they prepare to take the plunge into the Amazon basin, a basin a thousand miles across and a plunge of 11,500 feet in 20 miles. There were no reliable maps of the most difficult part of our proposed route, just a dyeline print we managed to procure in Quito; only rivers were marked on this otherwise blank sheet.

People often ask why it appears to be the same adventurous bands that go on expeditions year after year — a kind of old boys' fraternity. Certainly, virgin peak baggers and jungle bashers of Britain are a close brotherhood whose names creep into the national press with the monotony of obituary notices. But there's sound reason for this closed shop: they simply prefer to enter into dangerous situations with people they know. It's common survival sense to tolerate Lanier MacHete's habit of wiping dixies with his dirty underpants provided he won't abandon you when the chips are down (or all eaten). In this respect it was natural for Joe to suggest Mo Anthoine and Martin Boysen. As far as I have observed, they don't misuse their underpants and definitely wouldn't leave you in the shit. Mo was a key member of our Lost World expedition in 1972. This was a 'vacation' into the jungles of Guyana whose objective was to force a line up the overhanging scorpion-ridden prow of Mount Roraima for the peace of mind of the government of Guyana (who had never set foot on this far-flung outpost of their territory) and for the adventurous appetites of the BBC and the *Observer*.

Martin Boysen, tall, laconic and as competent on vertical rock as in dense bush, professed enthusiasm in this proposed preamble from the Llanganati to the Rio Napo, a tributary of the Amazon. Martin, who is a naturalist, has an academic interest in all things great and small and shares our liking for places far from the madding crowd.

Recalling the 'tired' locals from the last trip, I suggested to the others that we should try to get forest Indians for the journey. Our recollections of these natives from Guyana were those of an average motorist's view of a Ferrari — a rapidly disappearing one. They could glide fairy-like through the forest with 100 lb. on their backs, leaving us to bumble along like OAPs. Alas, this plan was not to be. It proved too difficult to procure them and anyhow the night frosts of the Llanganati would not have been popular with men used to 90 degrees Fahrenheit in steamy Amazonian shade.

I discovered that it is difficult to keep to a schedule on a standby ticket, especially when about 100,000 Colombians were returning from the USA to their respective folds for Christmas. The result was that I couldn't get out to Ecuador before Christmas. However, by the 5th of January Santa Claus had folded his sack for another twelve months. The airlines were quiet and five of us had arrived in Quito to meet up with Brian Warmington.

It was not exactly unexpected that my application to the Government of Ecuador to visit the Amazon basin should have gone missing in the British diplomatic bag. In any case, there is always something to go awry on any worthwhile expedition. If it had not been this particular mishap, it would have been the wettest season for 200 years, or there would have been a cataclysmic earthquake. Through the good offices of the British Embassy and the patience and helpfulness of the Ecuadorian authorities, I got the required permission in a couple of days with Brian having to sign his very soul away as a guarantor of our good behaviour. It was apparent that he didn't know us very well!

The reason for the difficulties which we hadn't encountered the previous year was partly to do with the offending television film, but also because the son of a South African farmer had attempted to travel from the east side of Cotopaxi south and east to reach the Rio Napo. He and the two Indians with him lost all their equipment over a cliff and were without food for fourteen days. As a last resort he prayed and this request received top priority, for they then stumbled upon a native shack and a bag of sugar. Later they were rescued by helicopter by members of the Special Services, a military commando-type group. I understand that this young man has since become deeply religious and I must say it made me think twice about this prayer business. The area we hoped to traverse was probably worse than the ground east of Cotopaxi, largely unmapped, with very steep gorges and escarpments and no trails. We had no idea whether we could complete the journey: I had simply spotted the possibility of such an excursion on the sketchy map of the Orient region of Llanganati, a map incomplete enough to stir an arthritic-like tremor in my bones. I realistically acknowledged that it certainly would be wet enough to ensure future attention from that malady.

There is something truly fascinating about the Amazons (for this river system is often referred to in the plural) — they hypnotize one like a great snake. The true source rises in Peru, only 120 miles from the Pacific, and winds the 4,000 miles to the Atlantic. It is 200 miles wide at the mouth, spewing forth one-fifth of all river water on earth. Even 100 miles upstream it is seven miles wide in places. Ocean liners can navigate up to 2,300 miles inland. The huge Amazon basin is no more than 650 feet above sea level with an area of 2,500,000 square miles. There are over 1,100 major tributaries and seventeen of these are over 1,000 miles long. And so it goes on: everything about the place is staggering. There are 1,500 species of fish alone, enough to keep Joe Brown and Yvon Chouinard happy for the rest of their days. The earliest European came in long before our landfall: Francisco d'Orellana was the first to navigate the Amazon in 1541. It was he who first talked about those fair-skinned female warriors, but the river itself was discovered in 1500 by a Spanish captain, Vicenti Yanez Pinzon, who was seeking a route to the Indies. An expedition on a grand scale took place in the years 1637 and 1638. Pedro Teixeira travelled from Para, now Belen, to Quito, with 2,000 men and 47 boats. Not being content with this journey he returned downriver.

Brian's hotel in Baños was buzzing with activity as we packed food and equipment for the trip. Martin and Mo had been delegated the task of victuallers back home and Martin had sent me details of the extra food to be purchased in Quito. This list occupied half a postcard. I was even more concerned for our well-being when I saw the amount of food being taken from the UK: a single rucksack. . . . My mind, however, was put to rest in Quito when Mo started buying as if a famine was imminent. Mo has the uncanny ability to buy and sort out the food requirements of an expedition in minutes. Rarely is he wrong in quantity and this pending expedition bore testimony to his skill, for when we arrived at the end of our journey we had four freeze-dried meals left. I'm convinced that his true vocation is that of a caterer, not a manufacturer of ice axes and hard hats.

Owing to the wide range of temperature which we expected to encounter, we had to take warm climbing clothing for the first part of the journey and this we hoped to send back with some of the porters. Just before I left the UK I had a letter from Brian saying that he had met a Spanish geologist, who had been part way up the Rio

Mulatos — the river we hoped to follow to lead us down into the Rio Napo and the Amazon basin. He had gone up from near Puerto Napo, a small town on the Rio Napo, which is on the Amazonian 'ring road'.

He had told Brian that there was a trail on the south side of the Mulatos for a way at least, but it was both rough and difficult to follow. There was no way that boats could be used on the river; it was far too turbulent. He had procured the services of an Indian guide, Nelson Cerda, who lived at a collection of huts which had the inviting name of La Serena. This small village was on the banks of the Rio Jatunyasa, a continuation of the Rio Mulatos.

This was something in our favour at last. After being pushed around by the implementer of Sod's Law, Dame Luck seemed to smile on us, for Evelio Ferreira, the geologist, was now staying at the Hotel Sangay — Brian's hotel. Whilst the others were packing I quizzed him on the Mulatos trail until all self-respecting mosquitoes were tucked away in hammocks. He marked out the trail on our sketchy map as far as he could remember, but he thought our plan mad in the extreme and quite impossible. From past experience I have found such comments provide an excellent incentive to go and do the bloody thing.

Evelio spoke good English and I asked him, "What are the chances of hiring two forest Indians to meet us on the Mulatos trail, once we get through from the Llanganati?"

He thought for a moment. "There's no communication with Nelson Cerda at his village, but you could possibly send a message via the Catholic Mission, across the river. The best way," he continued, "is to pay for transmission time on the Amazonian commercial radio station at Tena. Such a transmission should be picked up by the missionaries and relayed by runner to La Serena, where Cerda lives."

"How much would this air time cost?" I asked with Scottish caution, thinking of commercial radio tariffs back home.

"There's a charge of twenty sucres for five separate transmissions," he told me (about 40p).

"I don't think that should bust the sucrebag," I laughed. "How about sending someone from the hotel to Tena with the text for the broadcast, Brian, and advance payment for two Indians?"

"Sure, I'll write it out in Spanish," Brian said. "What shall we say?"

"You can use my name if you wish," Evelio generously offered. "It will lend weight to the request as I hire Indian labour for the Government when making surveys."

"Smashing, Evelio," I returned. "How about something like this, Brian? . . . 'Two men are requested to travel up the Rio Mulatos, one day beyond the Rio Yarepa, and meet a British expedition travelling from the Llanganati. Please wait there from the twenty-first to the twenty-fourth of January. Part fee of a thousand sucres payable at the radio station. Evelio Ferreira.'"

By now we were resigned to taking the Parama Indians and the day before Brian and I had gone up to San Jose di Paolo, the small wild-west town on the uplands above Ambato. Geraldo Angradi was the local leader and labour organizer, and when we met him he had been celebrating; the doubts which we voiced were swept aside with alcoholic fervour. He introduced us to a stocky, rather truculent-looking Indian who owned a shotgun and who, we were assured, would get us much game on the expedition. As we had hardly seen a creature on our last visit to the Llanganati, I was somewhat dubious about this claim, but Dead-eye Dick — as I nicknamed the potential purveyor of fire-arms — was duly put on the payroll.

Three days later we were back in San Jose in a hired truck with all our food and equipment. I think it was Edward Whymper who once said that what the Ecuadorians need is a good hard winter to shake them out of their lethargy. Certainly the Indians on the high Parama don't bubble over with energy.

By the time we reached the end of the drivable road under a canopy of dust, it was early afternoon and piping hot. Dead-eye Dick and a self-appointed shop steward stated that it was too late to go further that day. They were obviously missing their siesta. Joe pointed out that he had fished at Yanacocha, a lake on our proposed route, during our last visit when I was down with sunstroke, and it had only taken them two and a half hours to reach the lake. So they changed their plea to another dimension, from time to temperature: it was too hot. Being used to such porter procrastination on other continents, we picked up our own rucksacks, which were probably heavier than those of the porters, and set off. Presently, behind us, came a squelch of welly boots, the standard footwear of the Llanganati and upper Amazon. The dissidents had conceded defeat, but were obviously nursing their wrath at ambient temperature.

Our plan was to take the porters as far as Cerro Torres, the most north-easterly peak of the Llanganati, and there pay most of them off. From this base we hoped to explore and, with the current soaring gold prices in mind, have a look for the old Valverde mine. Then, when the heavier food was consumed, push on, using freeze-dried food to complete the journey down to the Rio Napo. Geraldo had been told by Brian of these plans and claimed he had been to Cerro Torres once, but none of the others had; it was isolated and difficult to get to, as we later discovered. The porters failed to make Yanacocha that night so we had to camp out on the pass leading to it. It was fine, open country with rolling hills and sporadic clumps of arrow grass, a labyrinth of bogs in the valleys and isolated, strange-looking 'standard lamp' plants about twelve feet high. To the south Joe and I could pick out our route which followed a series of ridges running south and east to Cerro Hermoso, the highest peak of the Llanganati.

We unfolded our tents and a camp sprang to life, like a conjuring act on that vast sunlit stage. Two fires lazily sent smoke into the clear, still sky. After air, bus and, finally, truck travel, it was good to be away from mechanical objects, other than Dead-eye Dick's peashooter and cameras — those essential appendages of latter-day explorers. Joe Brown selected six tea bags from the master cache of 1,800 and Joe Reinhard witnessed the first ritual brewing of British expedition tea.

On our previous visit Joe Brown and I had met Brian at his hotel. Although he was keen to go into the Llanganati with us then, a foot injury prevented him from doing so, and on the strength of this brief encounter we had invited him to join the present happy band. As a fluent Spanish-speaker he would be invaluable, but we knew only too well that getting along with professional expedition types like us requires the tolerance of a Buddhist and a barrack-room sense of humour, not just a knowledge of the local lingo. In fact, the mail from Baños was so unreliable that we had no idea if he could accompany us until we met him on arrival. I think both Joe Reinhard (we decided to call him 'Rhino' to avoid confusion with Joe Brown) and Brian were taken aback by Mo. To Mo nothing is sacred; even at the height of a violent electrical storm he will defy the Almighty with a rare choice of invective to strike him down. When he acquired a church in the damp depths of the Western Highlands, he transformed the building into an ice-axe factory. Then, taking the pulpit to his pad in Nant Peris in North Wales, he installed this rostrum of God as a cocktail bar. From his encyclopaedic memory Mo has a quip and a tale for every occasion, a repertoire recruited in his wanderings over the globe. He's the only person I know who can use up all the pages in his passport in twelve months. That day he entertained us with advice from an Australian cobber of his who once urged him to use a rounded pebble between the cheeks of his arse to alleviate chapping from which he was suffering in the outback.

"It definitely works," Mo assured us earnestly, "but I was warned not to confuse it with the pebble I used under my tongue to allay thirst!"

That night a sliver of a moon rose upside down. It sure seemed a topsy-turvy world with a crazy moon and those standard lamp plants which, we were told, only flower once in their whole life-span.

The frost wrapped up the Llanganati in a shroud. I slept peacefully, but just before dawn I awoke to hear the long roar of a puma close by. I gave Brian a dig in the ribs and told him about it.

"It's probably a jet," he replied with a yawn. But his yawn was cut short by another roar, even closer this time. Later, this puma, or its close cousin, gave Geraldo and Dead-eye Dick a sleepless night.

The dawn was as I can imagine it depicted in Llanganati colour brochures 50 years from now. It had an almost hypnotic beauty. The sky was a luminous blue; a breeze from the Amazon tickled the heads of countless grasses.

After breakfast, Joe Brown went on ahead to have a cast in Yanacocha. He left his telltale prints on the mud alongside fresh puma pug marks. We followed later, passing through clumps of arrow grass and gnarled, stunted trees, and there below us was the lake, shimmering like a polished amethyst. On our right, to the south, rose Cerro Hermoso, elegant and shapely like a girl in a wide flowing green skirt. A vision of Silver Camp crowded my thoughts. Was old Brunner right, I wondered? Was the gold holed up there, or had the Corrigidor Puga Pastor found it? Was the Valverde mine another thing altogether? I felt we were now on the correct route of the old map, but would we go wrong like so many before us? We'd no doubt find out soon enough. Ahead was the rain forest and, as I knew, hardship. On the other side of Lake Yanacocha the young river Golpe girded its loins for its long descent to the Atlantic. It was to change its name many times before meeting the Big Mother Amazon.

We could now say goodbye to the Altoplano as we passed out of the bright sunlight into the shadows of the jungle. We had twelve porters, two to act as trail bashers when we ran out of existing trails. Meanwhile they carried small packs which wouldn't have overburdened an old lady on a frugal shopping expedition.

The path began to dip steeply and we dropped into the verdure of the Orient. The Indians were in good humour. It was obvious that they now felt they were on their annual picnic and were resolved to avoid overstraining themselves. When we came upon a lonely hut close to the junction of a river flowing in from the south, the porters decided that it would be our stopping place for the day, even though it was only 1 p.m. They had already prepared their defence. One: exhaustion, and two: there were no more clearings ahead, they said. Some of the more militant unions of Britain could pick up a few tricks from the porters of San Jose di Paolo! After a lot of argie-bargie and a show of solidarity on behalf of the expedition shareholders, two of the Indians, Geraldo and his cousin, volunteered to go on and recce the trail, if there was one. The sole occupant of the shack, a well-bitten native whose only ambition in life was to acquire one of our cooking pots, said that it stopped a short way ahead.

The problem with overnighting at any 'civilized' spot in the forest is that it also generally provides lodgings for numerous hungry parasites. In this instance, the first resident to introduce itself with its fangs was the voracious horse fly which, as there were obviously no horses in the vicinity, willingly adopted us. However, we put two tents up and Mo and Joe decided to sling their hammocks under the verandah of the shack. It was a verandah without a floor, for the Indian Robinson Crusoe had run out of either steam or wood, or both.

Later that day, whilst studying the old map of the Valverde treasure trails, I saw that it was here at the river that Henry Longo, a priest and one of the early gold diggers, was drowned when searching for the Valverde mine.

Joe returned from the river with several trout and immediately inspected the brew on the fire — it was to his satisfaction. As he fished the statutory six tea bags from the large kettle he threw them over his shoulder in gay abandon, using a fork like a siege catapult. The first two bags scored a direct hit on the face of a porter who was sitting behind us on the grass. Wiping off the offending brown liquid with the back of his hand, the Indian promptly moved away, looking slightly insulted. Joe, now appreciating the error of his ways, amidst ensuing laughter, altered the trajectory to what he thought was a safe vacant lot behind. It was, however, the newly appointed sanctuary of the porter, who was not amused at receiving a second salvo.

"Hey, Martin, Brian is covered in small biting insects!" I shouted across from my tent which I shared with the nonplussed host of these beasties.

"Nonsense, they're seeds," replied our science master soothingly. "I've got them all over my legs . . . "

"If they're bloody seeds," I retorted, "they can do about five miles an hour." It transpired that they were 'garapati', tick-like vermin which burrow under the skin. Their main failing was indecision. They appeared to be very selective, dithering about, deciding where to make their excavations. Their 'spadework' was very irritating, but their strength lay in numbers; there were hundreds of them and they made themselves comfortable in the least likely places. For example, I found one ensconced in my navel; it had made a snug nest in some pile from my Damart vest. I have since learned that a Russian professor who made a study of similar ticks discovered that they home in on radioactivity emitted by the brain. Had I known this at the time I would have suggested that Mo manufacture steel helmets rather than those of glassfibre for explorers.

Next day we scratched along the trail which ran close to the river. The weather was good and I was thankful for the research done into the best time of year to travel in the Upper Amazon rain forest. The forest Indians, I had learned, choose the month of January as the rivers are usually low. Also, at this time, the fruits of certain palms are available, which make long journeys less hazardous, or at least less empty bellied.

A few miles further on we came to the pathetic skeleton of a burnt-out hut: a victim, we were told, of arson by Indians hostile to the family living there. That day also the first snakes were spotted, junior members of their race, but they were a reminder of what we might expect downriver. Reconnaissance sorties of mosquitoes were spotted, too, and these doubtless spread the news of free meals to their mates.

Dead-eye Dick shot a Torrent duck as we stopped for a midday break. This was cooked immediately over an open fire, though Martin, a keen ornithologist, said he felt like a cannibal, sampling such a rare creature. There were, however, plenty of them and I loved watching their chuckling, high speed paddling up rapids as if they were late for work.

I tried panning for gold at this spot, using Dr Rhino's soup plate, but with no luck. The porters, who watched with interest, nodded knowingly; there were no colours or traces of alluvial gold. Everyone was now de-garapating like mad, even the porters. They paired off like monkeys delousing. Later they were to use this epidemic as an excuse for abandoning the expedition. We soon realized, from the depth of the Lost World expedition memories, that, despite the lianas, this was not a Tarzan and Jane arcadia. There were mutica flies and puim flies, not to mention the jigger fleas that burrow under one's skin and lay eggs. There was also a delightful fly which combined tactics with the 'mosy' and laid eggs via its pal's puncture holes. The eggs in due course turn to maggots. But we weren't to be offered all these treats at once, and continued to discover new delights at every camp.

We had now run out of map. The detailed aerial survey maps which covered most of the Llanganati had given way to our skeletal dyeline print, which showed only one or two rivers; even these, we discovered, were inaccurate. I felt like one of the 'world is flat' brigade who had stumbled over the perimeter. The Indians didn't know where we were, or at least they didn't know where the trail took off for Los Torros, our tentative objective. So we continued downriver, fording several side streams and one fair-sized river, crossed with tree trunks. Dead-eye Dick shot some flying snacks; one was a quail-like bird and two had dark plumage and, until they had their heads blown off, sounded like woodcocks. It was a long day of tripping and falling along an apology of a trail which looked as if it had been last used by Conquistadors. In the late afternoon we reached a further river coming in on the right. It appeared to start its life somewhere in the bowels of Los Torros (that is, if Los Torros was where we thought it was). We followed this river to a point where it is recruited by the Rio Golpe. From here, the Rio Golpe is promoted to the title of the Rio Parcayacu, a name which it holds until joining the Rio Mulatos further east.

Here the Indians had a difference of opinion with us, and even amongst themselves, as to where the mountain was. All we could see in front was an anonymous 'Green Peace' drapery of yawning gorges toothed with tall trees. But close by we came across the remains of a lean-to shelter thatched with leaves. Beside the lean-to was a clearing which would have taken a garden shed. It was here that we pitched three tents. It had been a hard day; initially, the porters had taken notice of our taunts and had shown their paces like elderly athletes, only to revert to their soporific ways later, after their unscheduled eruption of energy. Dead-eye Dick went off into the bush with his blunderbuss, whilst Joe Brown and various porters headed for the nearest pool. The rest of us unpacked, undressed and debugged, by which time a triumphant porter came back with a rainbow trout of about four lb. — much to Joe's consternation, as his bag comprised only half-pounders.

Martin, who at this early stage showed a preoccupation with fire, announced, "Let's have plenty of hot water," and immediately hoisted a pile of logs on to the blaze.

"Why?" Mo asked. "Are you expecting a birth?"

"How about gutting the fish, Dr Rhino?" Joe asked, handing over half a dozen trout strung on a baby vine.

"I've never done it before," Rhino protested.

"Never too late to learn," Martin retorted. "Simple, look. . . . " He thereby dextrously gutted and beheaded a trout with a few deft strokes of his French cooking knife.

"You should let that idle fisherman do it, Rhino," Mo advised. "He's so lazy back home that his daughter winds his watch."

"That's all the thanks I get for feeding you bastards," Jo protested.

"Did you see the fish that porter caught?" Mo asked him, laying on sarcasm as thick as cold treacle. "That was a real fish, not a ruddy minnow like these."

Later, Martin went trail-hunting downriver and found an exotic orchid. Joe, who had recently developed a passion for photography, trundled off to record it with his new macro lens, but Martin didn't have any luck in finding a trail on the Rio Parcayacu which was the way we wished to go, now that we couldn't locate Los Torros with any certainty. There just wasn't any sign of a trail, though there were a few bamboos cut at an angle by machete; old timers' visiting cards: someone had obviously been down this far in the dim and distant past. The Indians were no more successful at trail finding than us, so we asked Geraldo and his left- and right-hand men to start cutting a new path. At dusk they returned looking disgruntled. The jungle, they said, was very bad and the machetes which we had bought in Quito were not heavy enough. Also, Geraldo pointed out, there was a trail running up the tributary on which we were camped and some of the older porters recalled their fathers saying that this trail led to Los Torros. From there, with illogical 'reasoning', they thought it might be possible to make a descent to the Rio Parcayacu. This, we knew, was extremely wishful thinking and we told them that in the morning we would try to follow the Parcayacu come hell or high water. We would probably encounter both.

In the morning there was a Gandhi-like silent protest by the porters. Eventually, Joe Brown, Geraldo and I went on ahead whilst the others waited, perched like budgerigars on bamboos. We cut our way to a point on a ridge where we climbed a tree and from about 30 feet up obtained a view of about 100 feet ahead. Whilst clinging to a liana the size of a transatlantic cable, I was reminded of a painting I had seen as a boy of an Indian pointing out the distant Pacific to some diligent explorer.

"It looks like a contract job for Wimpeys, Joe," I muttered in disgust. "Let's get back. We'd be eligible for meals on wheels by the time we got through to the Rio Napo."

Indicating with the machete rather like a traffic policeman, Geraldo conveyed to us that we should now try the trail up the tributary. We reluctantly agreed and returned to the others who were either lounging in the bamboo or ankle-deep in mud. The porters were relieved; they obviously had no burning desire to blaze a trail through virgin forest.

The path up the tributary climbed steeply and presently Joe and I, who were up front, took to the stream bed. Soon we were boulder-hopping and doing short rock problems which didn't seem to go down a bundle with the porters. At our first stop Geraldo, doing a tracker's act of spiralling out in ever-increasing circles, found the trail. Once more we trudged in Indian file up this mud canal. Presently we came to a man-made clearing of the usual slash-and-burn variety which the forest Indians cultivate for a few years until the nutrients are depleted and the ground is then left to revert to its wild state. We found the naked structure of a native shelter only recently vacated and our porters felt that *gringos* (foreigners) had been here. Two long thin poles placed against the cliff above gave access to an apology of a cave which we didn't even think worth investigating.

In a nearby pool I washed a few colours, but it wasn't gold in workable quantities. A couple of hundred yards beyond, the trail stopped and the ground was so steep and dangerous that we knew it wasn't worth attempting to continue. Despite the porters' saying this trail would give us a lead to Cerro Torres and eventually down to the Parcayacu, we were travelling south-west when we wanted to go north-east. My faith in my navigation, poor as it is, was certainly stronger than my belief in trail tales of an Indian grandfather. We descended to a now somewhat crappy camp.

IV

Trout for Dinner

THE PORTERS WANTED to go home. They had never been so far downriver before and garapatis were still nibbling them. We asked three of them to stay and help trail-cut for a few more days. Geraldo, his cousin, and Dead-eye Dick agreed to do so.

Next morning we were up at first light with the humming-birds. The porters were anxious to be off. Brian, acting as paymaster general, had the unenviable task of handing out the sucres. As usual, the inevitable haggling over payment ensued. At last this was settled and each in turn shook hands with us as they departed. Bidding farewell to the oldest porter, who had been the fall guy of the group, I failed to notice that he carried one of our better machetes. We had told them specifically that they could only take one of the lighter ones, which we called the breadknife. It was about fifteen minutes later that I realized that the two best machetes were missing. Mo discovered that ten tins of meat had also disappeared. We alerted Geraldo and his henchmen to the missing tools, and with oaths they took off like Olympic sprinters. They were a long time coming back, for the porters, with a homing instinct and a guilty conscience, had grown wings. They were overtaken a long way upriver, but must have anticipated pursuit, for there was no sign of the missing machetes, though Geraldo went through all their gear. When our three men of the Parama returned they didn't have to tell us that they had failed in their quest. They were like cowed dogs. As leader of the motley crew, it was obviously a slap in the face for Geraldo and he knew that, without decent machetes, his employment would be curtailed.

After they had a snack we sent them off to start trail-cutting with one of the remaining good machetes and the breadknife. Joe Reinhard and Brian accompanied them, load carrying.

It was obvious that to trail-cut alongside the Rio Parcayacu was going to take on the aspect of a public-works contract as we had expected. Our three musketeers (one at least had a musket) took to the river-bed when stretches of the bouldery beach could be followed — an Amazonian motorway. The boulders were incredibly slippery when wet, having their own peculiar brand of viscous lichen. Joe Reinhard and Brian did their fair share of cutting when they were forced into the forest, despite carrying loads.

On one of the river 'freeway' sections, when ahead of the Indians, they surprised a tapir and were thankful that Dead-eye Dick wasn't with them, even though we could have done with fresh meat. Like the rest of us *gringos*, they felt that the animal kingdom should be left alone.

That evening, in what could be termed a conservationist's nightmare, Dead-eye Dick marched proudly into camp with another Torrent duck, which we had in the soup. Grudgingly we had to admit that it had a rare flavour.

I had a busy day shelter-building whilst Joe was fishing with Martin. Mo was busy sorting out food loads for the duration of the journey. In so doing, he discovered that besides the missing meat and machetes, salami and biscuits had gone westwards with the porters. He told us this news as he stirred the cook pot.

"There you are, Hamish, you long drip, here's your dinner." He handed me a plateful of stew. "Made specially for you, dear, with no onions."

"Ta, Mo, I always knew you had a soft spot for me. Brown takes a delight in ruining my meals with those dehydrated toenail clippings." I have, as you will have gathered, a phobia for onions, which on this particular expedition was unfortunate, as there were masses of them. I was tempted to commit them to the wide waters of the Parcayacu when no one was looking, but resisted the temptation with fortitude.

By now we were all sitting round the smoky fire tucking into Mo's stew and Joe's trout.

"Onions are the essence of good cooking," Martin spoke sonorously, glancing at me.

"Can I borrow yer sp-oon, Jimmy?" Joe Brown asked, addressing me in his mock-Scottish accent. "I've lost mine, mon." He gave his usual infectious laugh which sounds like a revving outboard.

"Aye," Mo cut in, taking up the Harry Lauder patter. "It takes a long spoon to sup wi' a Fifer, or a Glencoer."

"Why don't you two Welsh wankers stuff those boxes of dehy onions in your gullets and jump in the river?" I retorted.

Brian and Joe Rhino must have been at a loss with our habitual expedition language, but even they were eventually smitten by the invective infection. Mo, a carrier of this disease, invariably spreads it like a dose of crabs (or garapati) amongst his fellow expeditioners, but in all fairness, the seed of his cursing usually falls on fertile ground.

Joe Reinhard is the archetypal always-on-location anthropologist. Tall, lean and, like me, losing more hair than his head likes to admit; a pencil and notebook of waterproof paper always ready and the easy-going manner of a Buddhist monk make him an ideal expedition companion. On our last expedition together, looking for a hidden valley in the Himalayas, he had been in his element, conversing fluently in the local dialects. Here he felt inadequate. This particular jungle was new to him, though he had made many trips in remote parts of the Nepalese Terai and Eastern Nepal. Towards the end of the expedition, he confessed that, for once in his life, he felt that he couldn't contribute anything. But this wasn't true; he was an unfailing and uncomplaining carrier of an enormous pack which we called the Pantechnicon.

Brian was the picture of the upright Sahib straight from a Victorian set. He looked more English, or at least British, than the rest of our motley crew; always with a scrubbed-clean look, as if he had just emerged from a washing machine. He has a passion for washing and I can't imagine what he thought of me, his tent companion who, like the others, washed only when the urge was upon me, or perhaps I should say, when a layer of Amazonian mud was upon me. The rest of the British contingent were uncouth, unwashed, and irreligious.

Early next morning, I set off downriver with the three Indians. Initially, I found the going good with a dried-up tributary of the river making life easier than the forest 'by-passes'. At least this was the case when the boulders were dry as they were that morning. The weather pattern had been helpful so far on this trip, with little rain during the day, though it usually chucked it down at night. Having arrived at the furthest point of the previous day, we climbed up a long slippery log to gain the top of a bank and started work.

At the first windfall, Geraldo, who was cutting a fallen sapling with the big machete, was left holding the handle; the blade had snapped and was lodged deeply in the trunk. This proved the last straw for the porters; all their enthusiasm evaporated. There remained only the 'breadknife', for Joe Brown's kukri — his personal possession — had also suffered a broken handle, though back at camp Brian was attempting to repair it. I couldn't persuade the porters to carry on with the 'breadknife' — as far as they were concerned, it was like asking a rally driver to finish a race on a bicycle. Geraldo now talked of the big machetes he could buy back in San Jose: two-handed jobs. He imparted this information through a series of comic-act impressions made with the broken handle of the implement. I vaguely recalled the shop opposite our cheap hotel in Quito where we had purchased these machetes and wondered if there was some significance in the fact that the woman who sold them to us had only one arm.

"Well," I said aloud. "It's bloody simple: if we can't go on, we'll have to go back. Let's go, you buggers." I pointed upriver and with sudden alacrity, as if they'd just been wound up, they took to the trail. I didn't see them again until I got back to camp. I did meet Martin, Mo and Joe, however, and told them of this latest disaster. They decided to carry on and see how far they could get. I thought I'd better return to camp and make some sort of arrangement with Brian to buy new machetes.

"It looks as though Geraldo is willing to return to San Jose to find replacements, but Brian will have to organize it."

"Well, we'll head on, anyhow, Hamish," Joe said. "See you back at camp."

"Right, take care."

On the way back I saw an enormous earth-worm in the dried-up river bed, the sort of creature that should be on the payroll of Hammer Films. Extended, it was almost four feet long. I had seen some of its relations on Roraima, but none as big as this one.

Despite the fact that they only had the 'breadknife', Joe, Mo and Martin made good progress and reached a point on the river where it was possible to cross. Also they saw a pass to the north which looked relatively low. When they told us their news round the fire that evening, I was interested in the pass; it would take us into new territory and might save us a lot of cutting on the river route.

"Are you sure that this crossing is feasible?" I asked, always wary when water is more than knee-deep.

"A piece of piss, Hamish," Mo assured me. "But wait till you try the rock-wall variation Martin and Joe did. It's the first extreme traverse in the upper Amazon!"

"Yes, it was rather pleasant," Martin agreed, in his unassuming way, recalling swinging across an almost holdless slab.

"Have you managed to make our three musketeers understand about going back for the machetes?" I asked Brian as I sampled scalding ox-tail soup.

"Yes. Geraldo will buy three big machetes and have them honed. If he hires a truck to take him up to the end of the road in the Llanganati he could do the round trip in five days — at a sprint."

"Ask him to get a couple of bottles of whisky," Joe Brown suggested, ever thoughtful for our well-being.

"And some fags," Martin chimed in. "You got enough, Mo?"

"Aye, ta."

"Fags and drink," I muttered. "Decadence in the jungle. I'm glad I don't share a tent with either of you two." Knowing my aversion to cigarette smoke, Mo and Martin took a malicious delight in blowing smoke in my direction at every opportunity, so that they resembled primeval dragons. Joe Brown also has an aversion to cigarette smoke, after smoking 20 a day for the greater part of his life before giving up. Even travelling out on the plane we hadn't been free of this perpetual smoke screen for Mo, with fiendish plotting, had arranged for his seat — and Martin's — in the smoking section to be immediately behind ours in the non-smoking zone. The gaps between the seats belched columns of smoke projected by our friends.

Geraldo's marathon run for the replacement machetes was going to cost us a bomb, since he insisted on having someone return with him. A wise precaution in such country, we grudgingly admitted. After all, as Joe Rhino pointed out, £50 wasn't too desperate an amount to blow on the job, which included the three implements. So, after shaking hands, the porters set off at 6 a.m. the next morning, disappearing with the alacrity of frightened snakes. We were not the usual brand of *gringos* that they had taken into the Llanganati in the past. In the first place we had insisted on going into a virtually unknown part of the region, then beyond it; secondly we carried as much as they did and were probably all fitter and stronger than them and, I'm sure, more competent at trail-cutting. Heading downriver into the Amazon basin was as alien to them as taking a hand at poker would be for a Wee Free minister. They were obviously glad to be shot of us, as had been their henchmen, before we led them into mischief. Dead-eye Dick, who jestingly remarked a few days previously that his wife had been enquiring if he was pregnant — for he had a fine paunch — seemed especially pleased at the prospect of returning home. He had been giving the garapatis his undivided attention for four days and hadn't shot anything for an equal length of time. Joe Reinhard had developed an aversion to Dead-eye Dick, insisting, quite rightly, that this trigger-happy Indian carried nothing, shot nothing (of late), did nothing but ate everything. He and his sweet-toothed mates, for example, devoured 60 lb. of sugar in six days!

There was no way we could carry all the swag downriver to the crossing point in one journey, but we each stuffed our rucksacks with 50 to 60 lb. of assorted goodies and staggered down the trail. During the previous two days we had carried odd items of equipment, ropes and spare rucksacks, and left them at various points to collect *en route*, a rash policy, for such 'left luggage' inevitably gets left and you have to make a special journey back from the next camp to collect it.

At Martin's Amazonian Hand Traverse problem we took to the bush. At least Martin did, with a determined look, brandishing the breadknife as if about to confront a killer pan loaf. I won't try to describe his 'easy' alternative to the desperate traverse. Suffice it to say that it went vertically upwards, ascending parallel bamboo bars — some with aggressive spikes; then in the Grand Old Duke of York fashion we marched, or rather swung, King-Kong fashion, down to the river. Joe Reinhard, after inspecting this trail of madness, took to the river, partly by accident I must admit, from a slippery rock and, despite being in ballast with his barge-like Pantechnicon rucksack, succeeded in gaining the easy boulder beach beyond the Boysen *faux pas*.

The river crossing was, as Mo had accurately predicted, a piece of piss. Debris hanging like bleached washing on high rocks bore testimony to more turbulent times. Times I didn't dare contemplate, for if the river rose to such heights now it would probably put paid to our enterprise — and us as well. It was perhaps with this in mind that I casually suggested to Brian that night that we take an escape line from the shore of the Parcayacu up into the forest where we had made our camp on the north bank.

When we first arrived the campsite became a hive of activity. As we cleared boulders to make room for the tents on the sand and to build a shelter, Joe Brown telescoped his rod and thereby combined his third most popular pastime with hooking our daily 'bread'. Martin crouched over a number of fire-resistant sticks, blowing and cursing alternately as if engrossed in a magic rite. In the rain forest there is nothing more exasperating than persuading a fire to burn rather than to smoke. Martin is not the most patient expedition member and occasionally he would hurl the plate with which he was fanning the smoke into the blackened logs and explode with a chain reaction of expletives. We all did the same when we tried it — only Martin made fire-lighting his personal crusade. From the river bank an effervescent Brown chuckled like a hen and admonished our science teacher.

"Now, now, Martin dear, don't lose your temper. . . . "

Mo, who has no craving for trout, even Amazonian trout, kept up a persistent heckling. "Why don't you do something useful, Brown? I put your bloody tent up every fucking night and all you do is fish, fish and fish. Christ, you make me sick, murdering the innocent buggers; you're worse than that poxy fox-hunting brigade." As Mo won no response to this tongue-in-cheek tirade, he recrossed the river and about half an hour later returned with a huge rucksack which had been left in one of our never-never dumps upstream. A little thrush (at least it looked like a thrush, but it couldn't fly) came over and watched me working on the shelter — a frame of dead-straight Baden-Powell-dream-poles thatched with plastic. Usually, I put up the shelter close to where the squatting Martin performed the ritual cursing of the wet sticks. We had noted that when the weather was fair, the wind blew down from the parama and when adverse, it came up from the Amazon, but invariably, through an inter-union misunderstanding of the smokeless zone, the fire was always at the wrong end of the shelter so that the structure resembled a fish smoker — which it usually was by late evening when Joe returned with his catch. The thrush observed these events from a distance of about three feet. Occasionally, its attention would be diverted to take a side swipe at a worm or some other slimy goodie which took its fancy.

I addressed our man of fire who had just returned from foraging for wood. "This bird looks like a thrush, Martin." When he saw it his face lit up.

"Queer little blighter," was his observation. Joe Brown, like Martin, was fascinated with Mr Thrush and indeed by all the wildlife of the jungle. Earlier that day, when I had spotted another giant worm, Joe was like a kid who had been given a mini computer for his birthday and insisted on Martin's holding it up for a photograph.

As we drank our last brew of tea that night, I commented, "Getting rid of the last of the porters is like having an enema."

"But I wonder if we shall ever see Geraldo again?" Joe Reinhard pondered gloomily.

"I doubt if he'll come down this far," I replied. We were now a hard day's trek from the last camp. We had decided to call this one 'Trout Camp', as Joe Brown seemed to be having unusual luck in murdering these creatures. The Torrent ducks were still with us, but otherwise there wasn't a great deal of animal life at this camp. Even 'mosies' were scarce and it was difficult to appreciate that we were on the fringe of the dreaded Amazonian basin, renowned for its profusion of creepy-crawlies. Only the vegetation bore positive witness to our location. It was dense, profuse and, once away from the river's edge, we were met by this tangle of nature's defences amidst trees as straight as Guardsmen.

Next morning Martin and I took to the bush. We were going to attempt to reach the col whilst Mo and Joe Brown moved downriver. The Llanberis contingent favoured the river edge as a means of reaching the Rio Mulatos, thinking my affection for the col somewhat misplaced. Upon our return that evening we were to compare unbiased notes on the day's activities. We left camp in the care of Brian and Joe Reinhard. There was always plenty to do at base, from collecting firewood to gutting trout and washing clothes so dirty that they would have broken the heart of an Irish drain digger, let alone a washer woman.

"Shall we start here, Martin?" I asked. We were standing on the edge of the river, and above the bank a uniform green barrier confronted us.

"It's as good as anywhere," he replied. "We've certainly chosen the day for this lark!" It was raining steadily as if trying to catch up with drought arrears. Martin had Joe's kukri and I had the breadknife. It is common practice in rain forest for two people to cut together; one does the rough initial hacking and selects the route, whilst the other enlarges the trail. The breadknife was ideal for this tidying-up operation and we worked well together. For a time it was easy; we were in the big timber where there was little undergrowth, and, indeed, travelling in the 'big sticks' can be pleasant; I'm always reminded of Thoreau's Walden Woods. But that day such idylls were short-lived, for we were soon back in the entanglements.

Martin was drinking at a small stream when he called me over. I was blazing a tree — a signpost for the unobservant in the jungle.

"What do you think of this, Hamish?" He held out a handful of coarse sand which visibly glistened.

"Pyrites?" I suggested.

"No doubt, but there is a fantastic amount of it."

"Sure is," I agreed. "I wonder if some of it could be gold dust? If so, we could give up this way of life for good.

"Let's take some samples back to camp when we return," I suggested.

"I'll wash them in Joe Rhino's soup plate." This was now established as the expedition gold pan.

"Good idea."

One of the few pleasures of manual labour is that you can let your thoughts wander whilst you operate on automatic pilot. At least that's the case when there isn't any danger. Whilst we were hacking away at this green inferno of frustration, I reflected on the near annihilation of our pet thrush at breakfast time that morning. As usual, the fire was behaving as if the wood consisted of soggy blotting paper. I was resuscitating it with a plastic plate which was vibrating like the fan of an agitated geisha girl. I had placed a large dixie of lukewarm porridge on the log beside me. All this was being observed in the grim grey dawn by Mr Thrush. Head sunk down on his shoulders so that he resembled a speckled egg or perhaps Humpty Dumpty in tweeds, he was about one foot from my right boot. Obviously he disliked rain more than smoke. Suddenly, with the vibration of my fan, the dixie fell off the log and successfully extinguished the smoke signal, at the same time smothering Mr Thrush with the gooey liquid. He shook himself indignantly, regarded me with a jaundiced eye — the other being covered with porridge — and stalked out into the downpour with what dignity he could muster. He never returned to our camp after that. As Mo dryly observed, "Your porridge must have been too lumpy for him, Hamish!"

Back in the jungle, such idle thoughts were rudely interrupted as I was slashed in the face by a springy sapling inadvertently released by Martin in his frenzy of kukri wielding. As any traveller in these hostile regions knows, a large part of one's time is spent in physical combat with the vegetation. Let me elaborate: take a common or garden path through a dense wood, then tilt it about 50 degrees and multiply the growth by 100. Now visualize the ground criss-crossed by a network of slippery roots sometimes covered in large slimy leaves like a camouflaged game trap. Above this level lie the first of the trip lianas, anchored at both ends, thereby forming snares so strong that only the machete or a pair of secateurs will sever them. Above this again come their compatriots, intermediate lianas and briar-like vines with barbs, ten times as big and many times stronger than the genteel English bramble. These are but the poor relations of the bamboo thorn, which evil device of nature I consider to be king of rain-forest armoury. Long 'runners' sprout from the trunk of this bad-tempered type of bamboo, bristling with thorns at eighteen-inch intervals. The thorns are over one inch long, possessing tungsten-like tips which could lacerate a bullet-proof vest. Not only are these barbed-wire tendrils devils to cut, but they reach over 20 feet from the parent tree. In this Wappenshaw of nature there are other bamboos whose trunks sprout vicious spines which at least ensure that your hand won't slip if you should grasp one. Some of these obviously have a defence pact with the 'upholstery' kingdom, for often the trunks are suitably clothed in a cuddly jacket of lichen or moss which makes detection of the spikes in no way certain until you grab the trunk. However, you get wise to these and caress the trunk, like a girl's arm, until you're sure that there's going to be no retaliation.

By far the greatest hazard in this environment is not the deadly bushmaster, or the corral snake, or the snap-happy piranhas. It is the severed bamboo sapling. When making a trail, the easiest way of slashing bamboo is to slice it off at 45 degrees some ten inches above the forest floor. This then provides the missing part of the animal trap, for the pointed stakes wait for the trip-vines to do their initial work. If the spiky palm doesn't fulfil its task in arresting your fall, the cut bamboo ensures that you will be securely impaled as a last resort. The spikes will even penetrate the sole of a light boot. Other ingredients which determine the discomfort of any budding Fawcett include windfalls, where you may have to crawl tortoise-like beneath fallen trees with a large rucksack on your back. Invariably the pack snags on each projection and drags on every vine. If you are foolish enough to sit down and rest you will inevitably provide public transport for a host of grateful insects, whose bites have a severity out of all proportion to their size. Subsequently, some of the bigger ants — over one inch long — have an effect similar to a four-inch nail being driven through your skin. And rain forests live up to their reputation. You are perpetually soaked, or at least soggy and muddy from the knees down. All this aquatic stuff is not to be outdone by sweat — the dinner gong for all those voracious little chaps of the forest. Despite these small points of criticism, the rain forest isn't too bad.

Such were the trials that Martin and I endured that day. Despite the steep angle of the slope, there were few places where we could get a view; as a matter of fact there was only one place where we could glimpse the river and that was from high up. The rain continued to hammer on the canopy high above us, but it makes little difference at this latitude being wet or dry; the rain was warm and it transformed our path into a treacherous runnel of liquid chocolate.

Martin has been described as probably the most powerful rock climber of his generation, but his efficiency overflows into other spheres. It was obvious to me that day that he had an eye for picking up a possible route through an impasse — possibly it's a similar gift to working out a line of weakness on a rock face. Martin has a strong passion for the West Highlands of Scotland where he has been a regular visitor, poaching rock-climbing plums on remote crags which few Sassenachs knew existed. Slashing away in my usual oblivion, I recalled a few years previously when Martin and I had joined Ian Clough on holiday in the Outer Hebrides. On the way back home we visited Blaven on the Isle of Skye where there was a climb that Ian and I wished to do. The locations of good new routes are always closely guarded secrets by the 'first ascent' fraternity, so Ian and I were a bit loath to share our 'plum'. Nevertheless, we eventually climbed it and Martin proved himself admirably — he made hard moves look shockingly easy. Ian and I had a further 'secret crag' which we wanted to 'clean up' at our leisure. Flushed with success, no doubt, we inadvertently let slip that this cliff was not 40 miles distant from the top of Blaven. That proved to be enough information for Martin. As a keen ornithologist, he carries around a fine pair of Zeiss binoculars. After Ian and I left (telling a white lie that we were returning to Glencoe) Martin did some spying from a suitable summit in the company of a good map. So that when Ian and I arrived later at Achnashellach, the stepping-off point for our 'secret crag', we were greeted by Martin who had already pitched his tent on the railway station platform (it isn't a busy station).

"Hullo, lads," he greeted us, "you took a long time."

"Yes, we had to go back to Glencoe, as we told you," I muttered lamely.

"I think there's an interesting crag up the back there." He pointed a long finger towards Fhuar-thuill. "It's called Mainreachan Buttress. It's about eight hundred feet high," he added with a twinkle, "and only one route on it, and that round the corner off the main face."

"Aye, we did hear a rumour that such a cliff existed," Ian admitted. "We meant to mention it to you."

This shrewd bit of detective work on Martin's part resulted in the climbs which we did being called after names derived from the detective industry: Sleuth, Snoopy, Investigator, Sherlock, Moriarty, etc.

Tripping over a root, I was wrenched back to reality.

"So much for Brown saying that this bloody col is a few hundred feet above the river," Martin spat in disgust as he did a pull-up on a root to gain the top of a muddy wall. "We must have come up over a thousand feet already!"

"Yes, I wonder where the col is?" I asked. "It was stupid not to have taken a bearing yesterday when we saw it; but surely we'll run out of trees and bamboo soon," I said without conviction. "It can't go up for ever. . . ." But by late afternoon it was still going up despite our ceaseless and laborious backing and climbing. Somewhere out behind us, where there was a gap like a skylight, the sun fingered through. We decided to pack it in. We were bushed — in the true meaning of the word. And Martin's hands were blistered. The slope was almost 70 degrees in places and the swathe which we had cut through it more resembled a gap between green curtains than a path up which to carry heavy rucksacks.

We rushed down this verdant gully. I held the breadknife at arm's length as if it was a stick of dynamite. I was so intent on avoiding being stabbed that I fell on to a blunt bamboo spike. Fortunately, I was wearing an anorak which prevented it stirring the remains of the porridge which I shared with the thrush that morning. Lower down we didn't forget our yellow dust sample from the stream; before I had a cup of tea back at camp, I was busy washing it in Joe's plate.

"Pyrites," I shouted over to Martin in disgust.

Mo and Joe Brown hadn't fared so well that day either. Their enthusiasm for the river route was shattered by a deep, dangerous gorge a mile or so from camp. They had to do some difficult climbing before finding it impassable. Cutting through the forest away from the gorge appeared to be a marathon task, too.

The next day Joe Brown volunteered to go with Brian on a compass bearing, using our path for a while at least. With this in mind Joe and I crossed the river a short way above camp and took a bearing on the col: 40 degrees magnetic. From here it was apparent that Martin and I had been some way above it to the west.

The weather had cleared into a superb evening; some parakeets gave an impromptu pop number in the trees and after the back-breaking day in the jungle everything now seemed in apple-pie order. We had a good meal of trout and rice washed down with 'tea of the bags', and as the wind blew downriver, Torrent ducks marked time on the rapids below us like Mississippi paddle boats.

That night the campfire talk took the usual random choice of subject matter, always providing a good opportunity for stories. Somehow we got round to the criminal fraternity and Joe Brown related how he and Chris Briggs of the Pen-y-Gwrwd Hotel in North Wales became involved in a court case concerning a wayward youth who had trespassed the law. After due consideration, they decided to do what they could to help the offender and both submitted separate letters to the court asking for clemency. Their plea had the desired effect and the transgressor got off. "But we were shattered," Joe said, "when he asked the court to take a further thirty-four offences into consideration!"

Talking round the campfire, or yarning when waiting out a storm in a tent in the Himalayas, is very much part of the expedition scene. Ours was a good diverse group and therefore provided much potential for interesting conversation. With the exception of Joe Reinhard and Brian Warmington, the rest of us had been climbing and going on expeditions together for years. Consequently, we knew many of each other's stock stories by heart and turned to our new friends for entertainment. Since Dr Rhino was a certain known entity, having joined Joe Brown and myself on a Himalayan trip, it was now Brian's turn to take the limelight.

It transpired that Brian had been a designer for Philips Electric before he settled in Ecuador. He had met his wife, an Ecuadorian, in Europe and, growing disillusioned with industry and the futile quirks of management, had decided to give Ecuador a try. Her father had a hotel in Baños and offered it to them. Now they have built it up into a thriving business with a blossoming future; Ecuador is about to be discovered. It is a unique country with much to offer the adventurous holiday-maker. Baños is at the base of the volcano Tungaurahua, 5,033 metres high. From here the road descends into Amazonia, to Napo and Tena, then back west in a big sweep towards Quito, and is linked to the main Andean highway from Ambato.

On a trip such as ours (as in the services), language can deteriorate to a marked degree. Somehow this seems to alleviate hardships; giving as well as taking, I suppose. As Mo said that night, the 'trick cyclists' would have some ruddy stupid explanation for it, with an impressive label. During our meal, we discussed the motives for exploring and climbing. Here again Mo offered benign wisdom.

"The eggheads would say that we're attempting to get back to the womb when entering an unknown valley, or reaching for mother nature's tits climbing mountains. . . . Bloody idiots," he concluded, dismissing them summarily.

Mo's conception of justice and arbitration rests with lengths of spaghetti. If the course of action decided upon involves the undertaking of unpopular tasks, then long and short lengths of spaghetti determine the fate and the day's toils for each member. That evening Martin and Joe Reinhard drew the short lengths — they had earned the unenviable drudgery of going all the way back to the San Jose camp to collect the remainder of the food. Joe Brown and Brian had volunteered to get to the pass. Mo and I were to fight a rearguard action at Trout Camp. I certainly wasn't sorry for this break, as my stomach felt as though it had been the doormat at an elephants' convention.

After the rain of the previous day it was glorious to feel the sun and get our gear dried out. Mo completed his third reading of a book, for by this time such prolific readers as Joe and Mo had consumed all the literature that the expedition had to offer. I had a preference for dozing rather than providing a candle beacon for moths and 'mosies'. Anyhow, I have observed that the candle invariably dies a spluttering death in the middle of a crucial chapter.

Brian and Joe Brown took the trail that Martin and I had hacked out with so much sweat the day before, then angled off it before the long precarious climb. They spent most of that day slashing through dense bamboo, a frustrating and time-consuming task. But they didn't reach the pass. They returned to camp, bedraggled, hands like pin-cushions with splinters and imbedded thorns, an occupational hazard in this quarter. Brian had also fallen victim to the machete — or on the machete — and gashed his hand. It really needed stitching, but when I offered to operate, with our communal repair thread, he hurriedly declined.

Joe Reinhard had been making serious inroads into his own supply of thread each night, as, Penelope-like, he would sew up his ex-army combat trousers, which were voluminous in the mode of empire-building shorts. But his subsequent and repeated pleating and sewing of trouser legs, converting them from culverts to drainpipes, was undone each day by the rigours of the forest.

It was three determined aspirant explorers that left Trout Camp the following day. Mo, Joe Reinhard and I were resolved to reach the pass. I had spent some time the previous evening honing the breadknife and kukri; brandishing our weapons, we climbed up the green passage. I had previously hinted that it would be sensible to blaze trees on the trail, as valuable time can be lost in taking the numerous *faux pas* made by those who had pioneered the way. Now the trail resembled the work of a billboard erector turned bushman. However, such 'signposts' were more than welcome; they gave one a feeling of security. Too often, when following an Indian path — if indeed paths they can be called — a furtive broken twig or an angle-cut bamboo give the only hint that a fellow human had passed that way; one who was an expert in cover-up and contortion.

Mo had gone on ahead; he likes to be on his own in the forest. As Rhino and I climbed we enlarged the trail, at least Joe did, for I was porter, carrying ropes and equipment which we intended to leave at our furthest point. I took bearings periodically, not that I could see much, but being behind Joe I got an approximation of our direction; when I couldn't see him I could hear him slashing with the kukri. We soon decided that Joe Brown and Brian had been within a degree or so of the correct bearing the previous day. This was heartening and, by the time we caught up with Mo, we found him in combat with a wall of bamboo.

"How's it going, Mo?" I shouted.

"No' bad, Jimmy," he replied in his best Clydeside accent. "It's a wee bitty dense like the hairy legs o' a Highlander and ma sporran keeps getting in the way."

"Perhaps Dr Rhino will remove it with his knife," I suggested.

By midday we had reached the pass. Had the undergrowth been found in the hothouses at Kew it would have been a major attraction. Leaves of what we called the broadsword plant rose twelve feet from the crest. By cutting some of these with the machete, we had our first view of the other side.

"Look at that flat area, Mo." I pointed to a densely wooded valley far below which opened to reveal a long expanse of scree. "You could land a plane there."

"A bit rough, old fruit," he returned, "but it would make a good campsite."

We later discovered, to our cost, that the so-called campsite was further away than anticipated and lower than we bargained for. It transpired that it was the spoil of a monstrous landslide, seven kilometres long by two kilometres wide.

"Looks a piece of piss for a bit," Mo observed, pointing immediately below to a steep slope clad with only a small amount of vegetation which was eventually devoured by the lower jungle.

"Let's go," I suggested, and led on down the slope. Once in the forest again, Mo went into the van, but after cutting for three and a half hours he decided that his hands had had as much as they could stand and elected to head back to Trout Camp. Joe Rhino and I decided to continue, but as the kukri had just been rendered useless — minus a handle for the second time — we realized that our efforts would be ineffectual.

"We'll see you back at the ranch, Mo," I said, cutting myself a heavy stick with the breadknife. "This, Joe," I said, brandishing the shillelagh, "will act as a flail: these are hard times."

There comes a point in jungle bashing when you don't think it can get worse, then you find that it can and does. That descent — and we didn't complete it that day — was abominable. My guts were sore from the earlier bamboo collision and I was feeling spastic. My cudgel, though reasonably effective in a grandiose Friar Tuck sort of way, was useless on the smaller, sprightlier offspring of nature, so a multitude of small vines and trip wires punctuated our trail. The only redeeming factor of that afternoon was a small bird resembling a kiwi that I saw, blinking at me from beneath a rhubarb-like plant with leaves the size of umbrellas. Obviously, it didn't have much compassion for my welfare, for it waddled away into the anonymity of the undergrowth with a disdainful air.

While I took some photographs, Joe cut to a point where he thought we could reach the moraine patch with about another hour's work. Then we slowly made our way back up that slippery vine-grabbing trail. We were tired by the time we reached the pass and gratefully sat down on a fallen log. Cloud had sneaked in during the afternoon. We felt insulated up there, just a stir of wind coming up from the Amazon and those giant leaves all around. We talked of a desert crossing which we had been planning to do and, despite the rigours, I was glad to be on that remote pass. After all, there are few places on earth where you feel so isolated for, irrespective of those green prison bars, there is a great sense of freedom in the jungle. It is like a swinging factory where you can relegate yourself to the position of guest observer. Watching everything that's going on, yet not part of the enterprise. These thoughts ran through my mind as we headed down the steep slope. Our perspective changed: over the top of the see-saw, we jarred down the rut towards Trout Camp. After the present expedition Joe told me he hoped to join another party in Patagonia.

"What are you going to do after that, Joe — go back to Nepal?" He had been working in Nepal for several years.

"Could do. I'm just not sure, but I'll possibly go back home to the States, just to see how things are there; it's ages since I've been." Joe is the archetypal footloose American graduate, perhaps seeking Sadko, the Bird of Happiness. I hoped he would find it at the end of his interesting odysseys.

There was good news at Trout Camp. Geraldo and another Indian had come and gone, leaving the new machetes and two bottles of rough Ecuadorian whisky — even that was welcome. Brian and Joe Brown told us of their visit. We had not expected Geraldo to come all the way downriver to our camp: it was the furthest he had ever ventured, but he probably felt a pang of guilt at leaving us to the mercy of the unknown. He told Brian that when they left us on their way back, they had managed to get right up to Yanacocha Lake and slept that night beside a derelict native hut. They didn't sleep for long, however; Geraldo felt a tug at his blanket and, wondering if Dead-eye Dick was getting amorous, looked up to see a puma crouched over him. His subsequent yells disturbed the beast and his companions. Afterwards, Dead-eye Dick kept a night-long vigil with his trigger finger at the ready. The cat didn't return. Perhaps the thought which had run through my mind when I had heard a puma's deep roar that first morning in the Llanganati was not so far-fetched. I could have sworn that roar had an ominously hungry edge to it.

The meal seemed especially good that night: it was a feast of the Passover for us; at least we had crossed one particular pass, and now our armoury was up to strength with three large Collins' machetes. After enjoying a dessert of Boysen pancakes, I spent some time honing the tools, though they were ground as promised. Our tea was reinforced with whisky and for once the campfire didn't smoke. The moon was out, and seemed to play a silvery tune on the river.

"How about a big push over the top tomorrow, Hamish?" Mo suggested. "We all carry and Wanker Brown and me could stay overnight on that large campsite. The rest of you could return with the clobber the following day?"

"Okay by me," I said. "What about the rest of the Park Board?" I looked at Martin and Joe Reinhard.

"Fine," Joe Rhino replied.

"Sure," Martin responded, lazily luxuriating in another fag.

"Okay," Brian said.

Like a conjuror, Mo did his division-of-food act and as usual we were in our sleeping-bags by 6.30 p.m. We kept to the working hours of the birds and bees: down at dusk, up at dawn. At least I was usually up then, but the others seemed more reluctant to stir. Joe Brown and Mo even insisted on having tea in bed.

Owing to my bamboo-barged stomach, I had the luxury of starting with the lighter load. We made good time up our new highway to the pass.

"How about calling this the Pass of the Observer?" I suggested. We had stopped for a breather — Mo and Martin for fags. "After all, we couldn't have done it without them." I was referring to the *Observer* newspaper.

"That's better than the Pass of the Turd," Mo volunteered, as he drew deeply on his cigarette. "I had a crap over there yesterday," he pointed, by way of explanation. "Just where Rhino's sitting."

"That's very considerate of you, Mo." Joe Reinhard leapt up as if he had sat on a bushmaster.

I went on ahead, enlarging the scrubby trail of the previous day and soon I was at Joe Reinhard's furthest point. He had left a note: 'Suggest you go right here.'

So I did, but soon found myself in hazardous country, where leaves covered gaps over roots and branches.

Now you see me, now you don't. Martin caught me up and enlarged the trail with his machete. The going got desperate; I don't think I can recall jungle so bloody-minded. It was desperately steep, with abrupt drops into gulches, and at one point a hairy passage had to be negotiated across a rotten wall of vertical vegetation. We labelled this the Hinterstoisser Traverse. Joe and Mo left a rope here for the others. After a final puzzle (ladders would have been handy here, for I'm sure there were snakes), we literally dropped into the stream bed with a sigh of relief. Our trials had already nibbled well into the afternoon and, as other, saner people in Quito were thinking of knocking off work, we were still pursuing our descent to the landslide. We hadn't forgotten that four of us had to recross the pass before we could have trout for dinner (Joe Brown had left a supply in a quiet pool).

"We'll go down a bit further with you," I suggested to Joe, in tones which implied that we were doing him an enormous favour.

"Gee, thanks," he muttered, giving me a playful stab in the ribs with his snake stick.

"Let's go, then." Martin spoke, always eager to be off, and the whole party moved in single file down the steep stream bed. Though the way was awkward, it was a relief to be out of the bush and we soon lost height. Reaching a stream junction, I was getting a bit worried about the time and suggested that we definitely had to be getting back if we wanted our beauty sleep. Secretly, I abhorred the idea of descending the trail to Trout Camp in the dark; it was dangerous enough in daylight, or rather in dim forest light.

"Let's give the buggers another thirty minutes," Martin temporized, "otherwise they'll never get back up here for these rucksacks, and I for one don't want to carry two down from here tomorrow."

"It's a deal then," I agreed.

In half an hour we reached a rock pitch in the stream bed which necessitated abseiling. It was a convenient place for us to part company. So, after Joe and Mo had descended and we had lowered the rest of the gear, we bade them *au revoir*, good luck, and headed uphill to the pass. We knew that they didn't have far to go to reach more amenable ground, since according to Joe Rhino's altimeter we had descended to a point 1,500 feet lower than Trout Camp on this side of the pass.

Martin had gone ahead, but I stayed with Brian and Joe Reinhard as Brian was suffering from cramps. As they had grown no worse by the time we reached the pass, I dashed on — my resolution of carefully picking my way down the dangerous forest trail gone with the breeze from the *parano* — compelled by my *idée fixe* of a mug of tea and fried trout.

When I arrived back at camp, Martin, like me, admitted to running down from the Pass.

"Violating the law of the jungle," I muttered as I took a deep gulp from a mug of tea which he handed to me. "Thanks. One of these days we'll be impaled on a bamboo."

Brian didn't fare so well and in fact fell on the descent, but luckily didn't hurt himself much.

We abandoned Trout Camp early next morning, still tired, leaving the shelter standing with surplus food suspended in a kitbag inside. We wondered if some poor exhausted forest traveller would be grateful for this one day — provided a flash flood didn't snaffle it first.

We made good time over the pass and were soon engrossed with the obstacle course on the far side. At the bottom of the abseil we found that the other two had collected most of the equipment and food cached the previous day. The remainder just tipped the balance of our loads, converting them from barely tolerable to intolerable. Reaching the enormous boulder field was like emerging from the confines of a cramped dark-room. It felt wonderful to be free, away from liana, vines, spikes and barbs. Here was an expanse of open country, albeit stony, vaster than we had anticipated. I yelled with joy to the others who were just emerging from the green tunnel: "We're here! Come on, you bastards, a brew is needed!"

We found Mo and Joe's camp a short way down the moraine, which we now discovered was the debris from a daddy of a landslide; the whole side of a mountain, to the west of our pass, had slipped away, leaving an ugly scar. The area was swept by this huge mass of debris, many millions of tons. Mo's little blue tent, like a submarine conning tower, looked singularly out of place on this area of devastation.

"Look, Hamish, here's a note for you," Brian said, handing me a sheet of notepaper. I recognized Mo's scrawl:

10 a.m.: Dearest Hamish. The 'Space Station' is partially erected. Hope you're capable of adding the finishing touches. We've gone on ahead and taken the dehy food plus a rope. Back for 3–4, I hope. The breast plates are in the valley SW of here. Yours in constant pain, Julian Vincent Anthoine.

"What does he mean by the Space Station?" Brian asked, thankfully dropping his pack on the stones.

"That contraption over there," Joe Reinhard pointed. "The shelter."

"What a bloody mess he makes of constructional jobs," I muttered as I picked up the machete and went in search of straight saplings to complete the job.

Martin, harbouring his usual preoccupation with the god of fire and the even mightier god of tea, was already squatting to one side of Mo's skeletal framework.

We had things more or less ship-shape by the time we saw our two friends coming back to camp up the long drag of the landslide.

"How did you get on?" Brian shouted when they were within hailing distance.

"We got down to a bloody big river," Mo replied, wiping his brow. "Going to be a pig to cross. It runs into the other river which skirts the bottom of this pile of shit." He nodded at the landslide.

"Is it the Mulatos?" I asked. According to our apology for a map, the Mulatos should have been fairly close to hand.

"No, it comes in from the right," Joe answered, but concerned for his first priority, asked, "Have you got a brew on, Martin?"

"Of course."

"Good man."

"I wonder where the blazes we are?" Joe Reinhard mused.

"Hell knows," I echoed, "but I don't suppose it'll tell us. There's some consolation that all rivers in these few thousand square miles of shrubbery run into the Amazon."

"We'll come out at the Atlantic if nowhere else," Mo added philosophically.

"You should have seen the bloody great stick insect that was on Mo's hair, Hamish. It was about eight inches long!" Joe enthused.

"It must have fancied him," I replied caustically. "About the only thing that would."

"Wait till you go down there tomorrow, MacInnes, you diddy, you can lead the vertical jungle pitch. It's just like Scottish winter climbing on saturated heather!"

"And there's a bees' nest for good measure," Joe chipped in. "Mo got four or five stings."

"Yes, the wankers have a hideout in a palm leaf the size of a canoe," Mo confirmed.

"How do we get across the river?" Martin asked, following his usual practical train of thought.

"Don't know, may be able to wade, otherwise some sort of Tyrolean. Hamish may be able to do a lasso job on some boulder on the other side; he enjoys those wild-west tricks."

"More interesting than straight climbing," I agreed. "Anyhow," I continued, taking my share of dehy stew from Martin, "let's get down to the nitty-gritty. Who's going downriver to try and rendezvous with that brace of Indians that's supposed to be heading up the Rio Mulatos? Brian must, as expedition mouthpiece. Other volunteers?"

"I'll go," Mo offered. "We've got three days, haven't we, before they head back?"

"Yes, they're supposed to return on the twenty-fourth," I said, "provided they got our message."

"What an arrangement!" Joe commented. "We don't even know the name of the river where we're supposed to meet."

"It may not have a name," I argued, "but they do know it's one day's march from the Rio Yarepa, and that's supposed to be a fair-sized river which joins the Mulatos from the north."

"The rest of us can go down with Mo and Brian tomorrow, and then return here," Jo suggested. "At least we'll get some of the food and personal gear a bit further and eyeball the crossing."

Just then a superb humming-bird came like a streak of light from the edge of the forest and, slamming on its brakes, stopped about ten inches from my face. It peered inquisitively at me then, in true UFO fashion, shot off at right angles to its original flight path. It was a magnificent specimen, larger than the others we had seen to date, but only one of the 319 varieties that live in Amazonia.

"You know, there's a mineral in the rock down by the river," Joe Brown announced. "At the bottom of the landslide; it looks like gold."

"Is that right?" I replied. "We must have a look tomorrow, but I expect it's pyrites. The place seems to be made of pyrites. This is a real fools' gold expedition."

Joe Brown, pouring more than his share of gut-rot whisky into his tea, reflected, "Talking about pyrites, it's a hell of a pity that we didn't have a chance to have a good look for the Valverde treasure from the San Jose camp. I'm sure we could have found a way up to Los Torros if we'd tried hard enough."

"Oh, I guess we could," I agreed. "The trouble is that the only trail led a hundred and eighty degrees in the wrong direction for our journey. Hey, wait a minute," I said, "that bloody mine . . . " I grabbed my rucksack and whipped open the flap pocket where the expedition treasure literature was housed, and singled out a copy of the old Guzman map, which was salvaged by Spruce all those years ago. I studied it in the last flush of evening, just before we were plunged into darkness for another twelve hours.

"Yes, I thought so," I said excitedly. "It ties in, Joe! Have a look, fella."

He studied the map as I traced the route. "That pass, the Pass of the Observer, must have been used by the Incas, it fits perfectly, and it looks as if the mine could be somewhere in the upper part of that flipping landslide."

"Yes, it sure looks like it," Joe agreed. "I think we should go and have a shufti the day after tomorrow. We can have an off-day looking for Mo's golden breast plates."

"Be sure that you keep some for me, you cheap gold-diggers," Mo cut in, "and just let me have a look at that map, Sahib."

"It seems incredible," I remarked as we sat round the fire, "that when we had no hope and no particular inclination to search any more for the treasure, it could possibly be right here under our noses!"

We all agreed that the terrain corresponded very closely to the old map; even the description, which I re-read later, in the tent, by candlelight, seemed apt. I thought of that document which Jackie, Mo's wife, had translated, back in Glencoe, which stated that the treasure now resided in the Royal Bank of Scotland in Edinburgh. 'Well,' I thought, 'you can't win 'em all, but there may just be a few grains of dust left.'

The treasure was soon put out of our thoughts the following morning as we shouldered our packs and headed down the moraine for the first river. I went ahead with Martin and tried panning the stream which ran down through the landslide, as well as the main river which it joined lower down, but found no colours. We saw veins of a yellowish mineral in quartzite, but agreed that this — Joe's 'gold' — was probably copper, though it was difficult for us to say as we were not expert at identifying minerals in their raw state. The only thing I could determine for certain, when panning, was gold, for I had been prospecting in New Zealand years before and had not forgotten the art.

We followed the main river now which, as it reached the base of the landslide, took a valley running south-south-east. We traversed along its southerly bank. Not an easy task, but it was a foretaste of what was to come for the rest of the journey. Slippery rocks and one rock problem after another. We came across vertical, smooth cliffs falling into the river, so had to take to the bush, often for considerable distances, and as this part of the forest boasts the densest undergrowth in the world, that wasn't easy. It was on one such diversion that Mo and Joe had encountered the bees' nest on the previous day. Before we reached this point we came across fresh pug marks of a jaguar. Now, too, we saw tall, stately Helechos tree ferns, probably a Cyathea, with an urchin hair style of fronds and trunks as straight as city lamp poles. These, as Joe Brown pointed out, are illustrated on the Valverde map.

I fixed a rope on the vegetable pitch Mo had reserved for me near the junction of the rivers. It was, as he had described it, steep, slimy and unpleasant, like scaling the side of a monster compost heap. Mo, who was behind, pointed out the bees' nest. It was difficult to appreciate that it was a leaf in which they had made their home, for it really was the size of a large canoe.

At last we stood on the edge of that foaming river, the one which Mo and Joe had reached the previous day. Viewing our Rubicon, we displayed as much enthusiasm for crossing it as we would have had for investing in a knocking shop renowned for its clap.

"We'll get our feet wet," I pointed out grievingly to Mo, who had the previous night displayed reserved optimism in getting over this obstacle.

"Hey, Hamish," Martin called. "Come up here." Martin's tall figure was perched on top of a gargantuan granite boulder; and, with his one-time white hat, ragged trousers and rotting duvet jacket, he resembled a latter-day beachcomber. I went up to join him. The boulder was waist deep in the racing river.

"Yes, it looks wet," I observed with trepidation.

"How about lassoing that boulder over there?" Some 40 feet away was a rounded boulder. To me it looked as if any possible attempt at such a rodeo act would allow the rope to slip off just when one was suspended over the rapids, another Amazonian sick joke for the benefit of desperate explorers. Providentially, Mo short-circuited any further aspirations Martin might have had in this field.

"I think I can cross below," he shouted. "Bring the ropes; I'll give it a go."

When I saw where he proposed crossing, I realized that this wasn't my cup of tea. Joe Brown well knew my aversion to water in its many forms (except as ice), but especially to rivers in spate, though no doubt this one was simply jogging along at its daily rate of 20 knots; awaiting the next flood.

With Martin belaying, Mo lowered himself into the river, then started to cross, leaning on a healthy bamboo sapling. By keeping the rope taut at an angle Martin was able to take much of the strain; nevertheless it was a cool act on Mo's part and the spray didn't even put out his fag. Once over, he anchored the rope and we quickly rigged up a Tyrolean traverse from Martin's boulder. For the first two the crossing wasn't so bad, but by the time it was my turn I was partly submerged and dragged across like the survivor of a shipwreck on a breeches buoy, much to Joe's amusement. As he danced about taking photos on slippery boulders, he announced with glee that he would select the best shot for his mantelpiece.

Leaving the rope in place for our return journey we continued

The author at Silver Camp

Llanganati: negotiating a difficult jungle section

Left: Llanganati: forging through the arrow grass, Cerro Hermoso beyond

Right: Joe Brown crossing the Rio Negro

The approach to Cerro Hermoso from the west

Mo Anthoine on the Rio Mulatos (Paracuma)

Expedition members at the Rio Negro

downriver. There were a lot of birds around. Martin spotted a macaw at the crossing and I saw an aerobatic display of parrots. From time to time we could hear them chatting away high on the canopy like old dears at a whist-drive. I found out that the Amazonian mosquitoes drink at all hours (and there are at least 218 known species). These monsters, and some of them are really big, set up their drilling rigs day or night. On the river edge they weren't bad unless you stopped, but as soon as you had a flirtation with the forest they were there presenting arms, or probosces, in your honour.

I have mentioned elsewhere the playful characteristics of the lianas, how they grasp you like the sinewy fingers of a witch. Yet some of them can be the explorer's friend, for they contain water. About eighteen inches of a big liana holds a pint of water. But beware all who suffer from jungle thirst, for others contain deadly poison. The sap of some are used to poison fish; another, the lonchocarpus, for example, has sap that will blister your skin.

As there was still no sign of the Mulatos after a couple of miles, we decided to return. It was going to take us until late in the day to climb back up to Moraine Camp and we had that ruddy Tyrolean to recross. Beyond the point where we left Mo and Brian we were sure that the river opened out and the Mulatos was just around the corner.

It was late afternoon before we got back to Moraine Camp, but Joe and I still had time to fossock about in the lower part of the landslide to inspect his 'gold'.

We had a big meal that night, finishing off with steeped prunes. Joe Brown, recalling no doubt his Cotopaxi eruptions of the previous year, accurately predicted catastrophic repercussions.

That night there were masses of fireflies, like will o' the wisps; Brian, commenting on them, had his sentence eclipsed by a flash of wildfire.

"Yes, they are bright," Joe replied in a dead-pan tone.

It was another fine morning and after a leisurely breakfast of sardines and Boysen pancakes, we set off in search of the Valverde mine. Joe Reinhard stayed at camp, deciding that his bush trousers needed a four-day service.

The heat bounced off the bare scree as from the reflector of an electric fire. The going was amazingly easy, almost like a two-kilometre-wide rough road with a drainage ditch, a new water channel, gouging it. Joe was so confident that we had stumbled on the location of the Inca mine that he took his headlamp. I felt the same way, though I had a sinking feeling that the landslide would have obliterated all sign of the old workings. The landslide was obviously relatively recent; there was little fresh growth, and what had sprung up was at the kindergarten stage.

It was much further to the breakaway scar than we anticipated, but when we eventually looked up at the source of the landslide, we began to appreciate its scale. The whole side of the mountain had slipped away, possibly due to an earthquake, or heavy rains, who knows? Anyway, it certainly swept the valley and filled it with rubble about 80 metres deep. The valley continued on up to the right of the landslide and we followed this for a short way until we came to secondary growth, for a much earlier slide had also come down here, but the trees and bushes now growing over it were quite substantial. We could, however, see up to the head of the cirque, since the valley ran into a steep ridge, which formed the crest of a large basin. To our left this connected with the peak of the landslide.

It was an idyllic day, probably one of the best I have ever spent in rain forest country. We sat on a rock and shared a bar of chocolate, revelling in the piping hot sunshine, for it wasn't hot enough to be uncomfortable, at least when we weren't carrying anything.

"Well, what do you think?" I asked the others.

"Looks as if the mine must be buried, or swept away in the slide," Joe said, adjusting a red stuff bag on his head which made him look like a Russian chef as he fossicked in the scree. "Hey, look at this!" He held up a large gold-coloured nugget.

Martin scrutinized it and dampeningly announced 'Pyrites', but we all felt that there might be valuable bits of scrap metal lying about in this part of the world.

"Shall we go up this other corrie?" I suggested, looking unenthusiastically at the head of the cirque, mentally calculating the time and labour it would take to cut our way through. It wasn't dense, but we'd had enough of Amazonian road-making for a while.

"I don't know," Joe replied, rubbing his knees which had been troubling him — possibly rising damp. "What do you think, Martin?"

Our scientific adviser seemed eminently comfortable, lazing on a boulder, luxuriating in a cigarette.

"I'm quite happy sitting here."

We were all quite happy sitting there. Later, we were to regret our idleness, but if one had a preview of life it would be incredibly dull. We freewheeled back to camp, anticipating a cuppa. But Americans are not switched on to the vital necessity of a brew.

"That bloody Yank has let the fire go out," Martin observed with disgust. However, it was quickly revived, so Rhino returned to favour with our expedition Prometheus.

Joe Reinhard stirred himself and shouted over from his tent, "Did you get the breast plates, Hamish?"

"Couldn't carry them," I replied. "They're too heavy. Mo's going to be disappointed."

The next day was wet and the river fairies had soft-soaped the boulders. Our packs felt as if we were transporting the total ransom for Atahualpa, breast plates and all. Rhino's pack looked and felt particularly heavy. Joe Reinhard is a canny soul and dislikes waste. The result was that if we decided to leave, say, some tins of sardines, or a packet of stew for which we had no great love or need at that moment, he would secrete it in his voluminous packframe and, bless him, usually produce it when we were in dire need of sustenance. He also acquired some items of discarded clothing, which I in particular chucked away, so that the contents of his pantechnicon were a subject of considerable speculation.

We crossed the Tyrolean traverse for the third time and with the additional weight of 280 feet of saturated rope we continued on our weary way, slithering and sliding like drunks with enormous carry-outs.

Mo had cached some dehy food at the dump where we had left them two days before and when we got there, this too was added to our swag. Mo left a note at what had obviously been their campsite. The message was positioned in a cleft stick, reminiscent of Evelyn Waugh's *Scoop*.

I went on for a mile yesterday afternoon, no sign of Mulatos. Probably at least 3 miles. We've taken enough food for a week and will keep going till we contact the porters, they'll be sent up to you. All this is hypothetical as you'll probably catch us up tomorrow. Aye, Moses.

Where we had expected to find flatter country, round a bend, proved to be as steep as ever; we were getting worried — where was the Rio Mulatos?

"I wonder," I addressed Joe Brown as we ate some raisins for lunch beside a huge river boulder, "if this could possibly be the Mulatos itself?"

"You know," he replied, "I was thinking just that as we were coming through the last lot of bamboo. That map seems crazy."

"If it is," Martin chimed in, "that river we crossed by Tyrolean could be the Rio Parcayacu again, coming round in a big loop."

"But it still doesn't fit with that apology for a map," Rhino objected.

"Well, I don't care much what river it is," Joe Brown replied, standing up stiffly, as if he required penetrating oil in his joints. "I'm going to see if there are any trout in it for lunch." There were, and Martin soon had a fire going. The sun took off his coat and a pleasant warmth flooded over us. We ate the delectable fish sizzled on pointed sticks poised over the fire.

Later, when we were skirting a greasy rock above the river, Joe Brown, who was leading, slipped and fell in. He went right under, rucksack, fishing rod, hat and all, but managed to get free of his pack and swim with it to the rocks. It was possibly the only section of river which didn't have rapids and, though the current was strong, an eddy caused by rocks lower down checked the flow at this point, otherwise he would definitely have been the first to reach the Rio Napo. After the initial shock of the incident and the assurance that he was uninjured, it was time to laugh. He dribbled ashore like a repentant suicide, took off his limp thermowear and we helped him wring it out. Joe Reinhard operated the other end of a contra-rotating wringer and the rucksack was capsized to remove part of the Rio Mulatos (or whatever river the water belonged to). It had been a sobering incident and made us realize the risks we were running. One false move in the wrong place — and most places were 'wrong' — could be compared with stepping from a platform into the path of a tube train: the end result would have been equally drastic.

The further we went downriver, the more convinced we were that it was the Rio Mulatos we were following. Mo and Brian had come to that same conclusion. They had been moving fast in the hope of reaching the rendezvous with the Indians. They saw both tapir and snakes. The first of the large Taocas soldier ants, the kings of the jungle, made their appearance here; everything gets out of their way. We tried various fruits, which the monkeys had been nibbling at before we disturbed them, but none of these tasted very palatable, and we were apprehensive about having more than a taste — I would have preferred a tin of Bartlett pears any day. We had been told that the sap of a particular tree, the Copali, could be used as a firelighter (it exudes a highly flammable resin), but we were unable to identify it until we had almost arrived at the journey's end. Had we managed to find this tree, it would have saved Martin much fanning and frustration.

Each morning the tents were soaking with condensation, rain, or both — which made a wearisome load to carry. As we tended to camp between boulders by the river, rather than in the draperies of the insect kingdom, we usually had a considerable amount of wet sand added to our burden.

It was late that day when we put the tents up; we were exhausted, but such is the resilience of the human body that by morning we were refreshed and ready to return to that seemingly interminable boulder-hopping treadmill. Often we were waist deep in water, then five minutes later hacking through dense bush. Fortunately, with Mo and Brian forging the trail ahead, this facet of punishment wasn't so onerous. We came across a camp the others had abandoned and there was the usual note. More optimistic this time because, as Mo stated in true Alexander Selkirk tradition, they had come across 'fresh foot-prints in the sand', which meant that the Indians had come up to that point. Also, we saw recent jaguar pug marks. At the edge of an old landslip, which spewed abruptly into the river, we came across a bamboo stuck between boulders, with another one inserted at right angles in the split at the top, obviously a Mo marker, indicating we should take to the bush.

We sweated up the edge of this fall and gained the forest, then found a trail that the others had made (and, presumably, the two Indians before them). Here we saw some very big bamboo, the biggest I had ever seen, like huge upended corrugated culverts, some with trunks two feet in diameter. There were also enormous rocket-like buttress trees. One I saw had buttresses rising fifteen feet up the huge trunk. Unfortunately, there was so little light that it was impossible to photograph it. Shortly after this, when I jumped down from a rocky shelf, a huge boulder crashed down behind me, missing me by a few inches. The subsequent smell of brimstone emphasized my lucky escape.

The task of following a rain-forest trail can be frustrating, especially when you lose it and have to back-track. This happened to us twice on this by-pass. At one stage we ground to a halt in a tangle of thorny bamboo. Eventually we did find the trail and arrived at another bouncing river. It announced itself long before we saw it, a deep roar like that tube train I was talking about. It was frightening and dirty. So far the rivers had been clear, with little suspended matter. This was the colour of mud and Martin, with his usual logic, observed: "I suppose this could be the Rio Negro?"

On one of the early sketch maps of the Rio Napo and Rio Mulatos, a tributary, the Rio Negro, was marked joining the latter from the south.

"It's certainly dirty enough," I agreed.

We moved down to the confluence with our own river. I'll call it the Mulatos now, for indeed it was. I was intrigued to watch the point where the two rivers merged, each clearly demarked for quite a way downstream by its distinctive colouring, the Mulatos turquoise green and contrasting like a crème de menthe cocktail with the muddy water of this violent newcomer. It was as if Mrs Black was engaged in a gigantic washing upriver.

"There's a stick marker down there," Joe pointed.

"And another on the opposite bank," Martin returned. "The others must have left them."

"Bold men," I said, eyeing the seething mass. Though we didn't realize it, the river had been much lower when they had waded across and Brian, who was ahead of Mo at the time, crossed even before Mo had caught him up. For us it was to be a risky business and I was grateful when Martin volunteered to go over first.

With one of our ropes Joe and I belayed him upstream and, using a pole as a 'third leg', he edged his way across. It would have been quite impossible without the rope and, indeed, our subsequent crossings after him on a rope, anchored on both banks, were equally perilous; if we had slipped we could have been pulled under — one of the dangers of using a rope for river crossing. I went over next, hauling yet another rope behind me, to enable us to retrieve them in the same way as when doing an abseil, by leaving a sling round a boulder and pulling one end of the doubled rope through it. However, this rope was not being belayed upstream as Martin's was, and was whisked away towards the main river in a great loop — feeling to me like a sea anchor, which inevitably snagged on boulders. There was no way I could yell to the others to free it; the roar of the rapids wasn't conducive to communication. All I gained from my vocal efforts were mouthfuls of water.

The only redeeming feature of running such an aquatic gauntlet is watching the subsequent antics of your friends when it is their turn. Joe Brown, being the smallest (Martin, Joe Reinhard and I are all over six feet), fared particularly badly and was lucky not to be pulled under.

As the day was well advanced, we decided to camp on the river edge and started to make our regulation shelter. Martin found a recently vacated camp, obviously constructed by the two scampering Indians. Hanging from a tree was an old pot with a small bag of porridge inside.

It started to rain before I got the shelter made, but getting wet didn't bother us; we had been soaked most of the day. We were now on strict rations of freeze-dried food. Though this is usually unappetizing, we were so ravenous that it tasted superb — what there was of it. Joe had no prospect of fishing this part of the river. There were no pools and it was too dirty: times were getting hard.

There were hundreds of butterflies, not the huge morpho butterflies we had seen in Guyana, the size of exercise books, but a wide variety of all colours, like an assortment of gigantic confetti. Joe Brown photographed one with BP on its wings.

"It's a bit out of place," I observed, "for it was Shell who did the original oil exploration in the upper Amazon. The company even parachuted pigs into the remote Shell Mera, one of the early prospecting camps. It's a small town now."

That day our route, which sometimes cut bends on the river, had rewarded us with displays of orchids, blood-red passion flowers and poisonous nightshade. It made me appreciate how Richard Spruce dedicated seventeen years of his life to studying the Amazonian flora. During his explorations, this self-taught scientist collected 7,000 important specimens. Everything is jazzy and larger than life in this jungle. Each species of orchid or flower seems to shout "I'm the greatest", and it probably is! Over half the world's total population of 8,600 tree species are here and, like the mosses, there are probably many more still to be identified. On the opposite bank of the Mulatos I could hear monkeys chattering in an avocado tree.

In the low light of evening the independent colours of the rivers as they merged were even more conspicuous; they didn't turn a uniform wolf-grey until a hundred yards below our camp. This, in comparison to the enormity of the lower Amazon, was but the widdling of a tiny tot into the Thames. As a matter of fact the Amazon pours as much water into the Atlantic in a day as the Thames manages in a year. It is not unusual for the banks of the Amazon to be under 40 feet of water, with the floods extending 60 miles inland from each side. The equator passes through the mouth of the Amazon so that the seasons, wet and dry, are divided. To the north the height of the rainy season is June,

and to the south, December or January. With this change the river rises
and subsides north to south so that it 'pulsates' like a huge heart.

It rained most of that night and the following morning. The sky
(what we could see of it above the river) was a battleship grey. I teased
some gossiping parakeets by blowing on my whistle; then we folded
our wet, sand-encrusted tents and staggered off downriver. It was the
usual steeplechase of boulder and jungle. After climbing up through a
chimney in the rocks, I found Joe in a most unusual position, lying face
down with his head almost in the river. I assumed at first he must be
having a drink, but thought this was rather an unusual way to go about
it with 50 lb. on his back, although his rucksack was not on his back
where any well-behaved rucksack should be, but was perched on his
neck.

"Are you thirsty?" I asked in all seriousness. All I got was indignant
spluttering as he did a painful-looking push-up. "Oh, I see . . . you've
had an accident!"

"You bloody long drip," he expostulated when he snatched some air.
"Have you ever seen anybody having a drink with a bloody great
rucksack on his head?"

"It did strike me as unusual, I must admit. Did you fall?"

I realized that this was not a diplomatic thing to say to the world's
best-known rock climber. We moved on in silence. . . .

We were having to be Scrooge-like with food now and our supply of
a lunchtime handful of raisins was finished. There wasn't even fish.
Though Joe had a good eye for a river, he didn't waste energy when
there were no pools or suitable eddies for the fish to loiter in and pick up
titbits.

Then we met Brian. He was standing on the wet rock beside the river
dressed in a fluorescent anorak, a machete in hand. We were glad to see
each other again. It was a sort of 'Dr Reinhard, I presume' encounter.
We knew that there was no news of the Indians, for had they managed a
rendezvous, Mo and Brian would have dispatched them upriver to us.

"We're camped about two hours downstream," he told us. "Mo has
gone on ahead trail-cutting because we have to go back into the jungle
again below our camp." He told us the rest of the news, saying that they
must have missed the Indians by only a few hours as their prints were
fresh in the sand.

I spoke with Brian as we clambered over those greasy rocks. We
talked of the loneliness of the jungle — something he had felt acutely as
he waited for us. I knew what he meant, for, like the others, I sometimes

went off by myself, either first or last on a trail; it's probably a similar desire to that experienced by the solo long-distance yachtsman. Despite the hundreds of varieties of trees and the multitude of insect life, there is a uniformity about this rain forest similar to the ocean. It is so vast, a 2·2-million-square-mile green sea; alas, already over 50,000 square miles have been annexed by man.

Mo was in camp with bad news. He had strained his back and was in agony. It had happened as he was returning from his furthest point on the trail, when swinging off a branch. I dug out some Fortral tablets from my communal first-aid and he gobbled a couple of these gratefully. It was painful even to watch him: we could do nothing. Martin did try a deep massage and Mo said that at least it gave him relief when the massage stopped.

The camp was on a high beach on a straight section of river, with a large pool at the lower end. Mo told us that beyond, where elfland-like bush-covered pinnacles rose from the far bank, there was another river coming in. This, we concluded, must be the Cotachi. Now things were beginning to make sense. So they should, I felt, for we must be well downriver now, close to the Rio Napo, a principal tributary of the Amazon.

As we dried our sodden clothes we resolved to avoid a decision on what to do next until the following morning, for, as Mo pointed out, he could make a rapid recovery. He had had this affliction before. Both Martin and Joe Brown were also 'back' experts, having suffered from various maladies of the spine over the years.

The morning brought an improvement in the weather but not in Mo. He was lying in his tent in a distressed condition as the rest of us squatted round the fire. I had built a lean-to shelter against a massive rock and the fire was crackling to itself alongside it.

"Let's sort out this dilemma," I suggested. "How about three going downriver and three staying here, two and Mo?"

"Might be better for just one to stay," Rhino advised. "It means those that stay could have more food."

"True," I observed.

"How much food do we need to get out, Hamish?" Joe asked.

"Enough for about three days, I think — if we're where I think we are! How many days' tucker are left, Martin?" I added.

"I would say a total of eight days' supply could be left for Mo and whoever's staying. That will leave about four days' bare rations for the rest of us, to get out. The first party will have to scamper like bats out of hell. Anyhow, let's draw spaghetti. You're exempt, Brian, we need your Español to arrange for Indians to come in to help Mo — should he need it."

I selected the spaghetti sticks with the regulation short one hidden in the throng. Martin and Joe Brown both picked long ones. "Where's Rhino?" I asked.

"Washing the dixies at the river," Brian reported.

"I'll draw for him," Joe said and drew the short straw. Rhino, on his return, thought we had worked a fast one (or short one). He had lost repeatedly at this game of democratic chance since the start of the trip and was now obviously put out; more, I thought, by the fact that his straw was drawn by proxy.

"Never mind, Rhino," Joe Brown said consolingly. "I'll stay. I'm used to Mo's cantankerous ways. He'd corrupt you and eat all the food."

"You can always put him down after a fortnight if the Indians don't turn up," I suggested.

There was now a frenzy of activity as we packed and divided food. We were ready to go by 9.30. Mo outlined how, on the previous day, he had come to a dead end on the trail some two hours from camp. He had tried numerous alternatives, but each seemed worse than the last.

"We'll leave cairns where we go into the forest from the river," I said. I was glad that I wasn't staying. Once a trip is almost over, as we felt this one was, you remember all the things you have to do back home.

Each succeeding day the trail became more obvious, in sections at least. We were left in no doubt as to where Mo had gone astray the day before. After a small stream where there was a colourful array of butterflies, we found paths radiating; the place resembled Spaghetti Junction. We stopped for a drink and to review the situation. "Hobson's choice," I muttered as I wiped my mouth. The cool water was delicious and it removed the lingering taste of the famine helping of cold rice which had been breakfast.

"Let's go down to the river," Martin suggested. We could hear it a short way off.

"Okay," I agreed. "We'll leave our packs here. You coming, Rhino, Brian?"

"I'll stay guard over the packs," the American replied.

"I'll go up this trail ahead and get some of my gear which Mo left yesterday. He said it was about fifteen minutes from here."

When Martin and I reached the river we saw that it was out of the question to follow it at its present high level. An old landslide fell steeply into the 'oggin and it looked too deep to wade. We later discovered that this was the way that our Indian relief party had gone.

Back at the Anthoine junction we tried several of Mo's diversions, mocked by a family of monkeys high in the canopy; they sounded as if they were pissing themselves at our dilemma. But we fared no better than Mo in forcing a route through the jungle. At one place we arrived in a veritable hell of thorns and rotten bamboo forming a lattice work over the alpine-angled forest floor, so that at any moment we could have been transposed from our present level to suspension on bamboo 'barbed wire' tendrils. But by now we were pretty determined and forced a way through.

Even in the agony of that day's march there were some precious moments when sunlit stretches of the river turned into a seething mass of silver with lianas draped like gossamer belays. Thinking about our arachnid friends, I was disappointed that there were no enormous specimens like the eight-eyed bird-eating jobs we had seen on Roraima. Some of those measured over ten inches across. But there was no shortage here of a racing model about the size of a saucer, which moved with amazing speed.

By now we were well into the Amazon basin; it was humid and the undergrowth was even more exotic. The forest floor was alive with the ubiquitous ants in their many forms. For example, the sauva ants, carried shield-like segments of leaf and moved relentlessly, legion-like. There was no slacking here, no industrial go-slow; just the imperative, resolute feeding of their mould chamber as sustenance for their young.

The day was preparing to put up its shutters before we stopped, shattered. As usual, we camped by the river to avoid open conflict with the animal kingdom in the bush. We were soaked, as was everything else, but Martin managed to produce a reasonable dehy stew after addressing the fire at length, expletives not deleted. We had four meals left, but were confident that we would reach some sort of civilization the next day or at least the day after. Martin was our trail detective, capable of recognizing the faintest clue of previous human passage: a broken twig, the suspicion of a footprint, or the healed slash of a machete on a tree trunk. Fortunately, the path was now more obvious — that is, once you found it — for there was still no warning when it left the river for the bush, and usually we had to go on to the impasse, normally rapids, a gorge or deep water, before back-tracking to inspect the forest fringe with Holmes-like thoroughness until the trail was located.

The following day was like any other, with its bum-steer paths. By afternoon we had climbed on a sinuous route up steep ground from a huge river pool, overhung by trees and epiphites, to find a twelve-foot-wide track. We knew then that we had come out at the other end. This was the edge of civilization, the extremity of a big trail from the Amazon basin and used, we learned later, by engineers assessing the potential of the Rio Verdeyacu for hydro-electric power. This river, which joins the Mulatos close to where we stood, contains gold, and higher to the north it is visited by a tough band of Indian prospectors who take a trail over an 8,000-foot-high pass, then descend into the bowels of a canyon of the Rio Verdeyacu. Just then the world was reeling with the spiralling price of gold and even a few miles downriver from where we were this demand was felt. All the Indian wifies were out with their gold dishes, panning for pin money to augment the family budget.

The new 'highway' rose steeply; we sweated madly. Suddenly we came across a large camp belonging to the workers employed in trail-making. Brian discovered from the cook and his young brother, the only occupants just then, that the track was being made to 'mule standard'. They allowed us to make a brew on their fire and after half an hour we bade them adieu. We had gleaned that it was some two hours back to the river at this point, for the trail went a long way inland. Now it started to ascend with a vengeance and continued to do so for what we estimated to be 3,000 feet. By the time we reached the final crest our legs felt 'shoogly' as the Scots say. Once over the crest, Martin and I lunged on ahead of Brian and Rhino. The descent was equally steep and strenuous, so it was an incredible bit of luck that we found somewhere to camp that night. The forest itself was uncompromisingly dense and there was no water, but a small clearing on a sandbank by the river proved ideal. Here the water charged through a parallel-sided gorge of solid rock with flood-water ablution pools. I made a bee-line for one of these and sat in it like a water buffalo, washing my clothes '*in situ*' and idly wondering if I was using gold as an abrasive to remove the grime. Martin, who always puts duty before personal comfort, busied himself in his interminable quest for dry wood and soon a band of smoke caressed the arms of the trees above the campsite. Presently I put the tent up and observed that it was drying satisfactorily after stewing all day in my rucksack. Shortly afterwards, a lithe Indian, dressed only in a ragged pair of trousers, sidled into camp, carrying a cloth bag and an old shotgun. He took a bunch of bananas, lemons and a packet of sugar out of the bag and offered these to me. He then produced a home-made cigarette from nowhere and presented it to Martin, divining that he smoked. It was made from home-grown tobacco and despite the fact that I'm a conscientious objector to cigarette smoke, I must confess that when Martin lit up I found the aroma delightful.

We were both touched by the spontaneous generosity of the hunter. Sugar was a great luxury to him, but he wanted no payment: it was simply a present. Obviously aware that we had come downriver, he no doubt knew what we had been through. It was with the greatest reluctance that he finally accepted the money I pressed on him. He was patently offended at first, but we insisted, as we didn't like taking advantage of his generosity — in many parts of the world it is too often abused.

The hunter returned that evening and Brian was able to speak with him. He explained the plight of Mo and how we required four Indians to go up the Mulatos in case he should still require assistance. The hunter had never been upriver, but he knew Nelson Cerda, the Indian who had come up to meet us. He confirmed that Nelson had done so and added that Nelson was the local acknowledged outback whiz-kid. Though the hunter volunteered to go upriver with his father, we thought it prudent to see Nelson first as the employment of the hunter could cause inter-bush union complications. Anyhow, we felt reasonably confident that Mo would get out under his own steam, and after all, they did have enough food for a few more days. But I asked the hunter (through Brian) if he and his father would carry our packs down to Nelson Cerda's as we were more than fed-up with our loads. He agreed.

Next day we reached the village of La Serena where Nelson hung out. We had to cross two more rivers to get there. The second one, just short of the village, was negotiated by dugout. The first thing I saw as we stepped ashore was a barbed-wire fence. "Must be civilization," I said to Joe Reinhard.

The village consisted of simple raised wood houses strung out at about 100-yard intervals as if the residents were anti-social. Nelson's was the most prosperous-looking. He and his cronies were in a state of profound intoxication when we arrived, with dilated pupils and reeking of a local gin. There was no way we could communicate with him; one of the womenfolk offered us a room, but it was so noisy with the gin pantomime that we pitched our tents in a nearby field.

Imagine our surprise when, in the early evening, Mo and Joe arrived at camp, Mo walking stiffly like a guardsman. As he had fervently hoped, his back had relented shortly after we left.

Next morning a red-eyed Nelson came over to camp, shamefaced and looking as if he had just served a protracted sentence on Cayenne. We met his wife who had a lovely miniature monkey, which clung to her hair. Nelson then told Brian, as an excuse for his blinder, that he had been in the forest for several weeks before he came up to meet us. He had our sympathy; we would have joined him if we had anything suitable to drink. The gin tasted like aviation fuel! We didn't press the fact that he'd left the rendezvous before the agreed time. There was no point in bringing that up now. Anyhow, it meant that we had completed the journey on our own with the exception of the initial frolic with the Paramo Indians. Nelson and the other Indians of the village who had joined us were impressed by our journey. They were, after all, the only people who could appreciate the difficulties we had experienced. Joe Reinhard asked Brian to find out what tribe the villagers came from. 'Catholic' was the prompt answer. The Indians of the Amazon basin have a lemming-like wish to lose their ancestral identity.

Through Brian I asked Nelson if he knew about the landslide. He did, so I asked if he knew when it had occurred. Though he wasn't sure, he thought possibly about eighteen months previously. He had only been that far upriver twice. I thought this fitted with the amount of fresh growth which we had seen. But imagine my surprise when he added that he knew of old Inca workings near the landslide. Like Brian, I was flabbergasted when he translated.

"Ask him to describe them, Brian," I said, trying to subdue my excitement.

Nelson told us it was a well-executed drystone structure several metres deep, part of it forming a channel with a circular basin at the bottom. "Where?" I almost shouted to Brian.

"Not far from the slip."

"Bloody hell, and we missed it, Brian. We must have been within a nugget's throw of the place."

"Hey, Sahib," I shouted to Joe. "Looks as if you were right to take your headlamp up to the top of the landslide. Nelson here tells us that that Inca furnace — at least it's probably that — was close by."

"Is that right?" Joe replied. "Well I'm buggered!"

"It could be the 'Guayra', the furnace which is mentioned in Valverde's description," I said eagerly to the others. "It all ties in. That report about Corregidor Antonio Pastor sending those treasure chests to the Royal Bank of Scotland in Edinburgh. He lived in Latacunda and Latacunda is the nearest village to the landslide, coming in from the upper Mulatos."

It all fitted. But did the wily Corregidor or Magistrate do the dirty on the King of Spain, I wondered, for it was the Corregidors of Tacunga and Ambato, towns close to Latacunda, that were ordered by the King with Cedula Real (Royal Warrant) to locate the cave and secure the treasure. Did Pastor subsequently transfer the treasure illegally to Peru for shipment? (For it would have been impossible for him to have disposed of it in Ecuador.) And where better an ultimate home for the treasure than with Spain's old enemy, England? (Foreigners, I have discovered, don't differentiate between Scotland and England.) As mentioned earlier, a map and description were bequeathed to the King by Valverde on his deathbed. "Ask Nelson if he has ever heard of Inca treasure in that region, Brian," I urged.

No, he knew nothing of the get-rich-quick map.

"Well, we win some and we lose some," I said despondently. "If only we'd made a more thorough search, Martin!"

There was no way we could even contemplate going back upriver to check this report. Yet the information excited me as it must have done the others.

"There go my breast plates," Mo said with resignation, "and I was so looking forward to polishing them. . . ."

We hired a couple of horses to carry our swag and, travelling light for only the second day in weeks, we headed downriver, down the Jatunyacu — the continuation of the Mulatos — to where dugouts were moored. On the far bank of the wide river we could see the Indian women panning gold. The previous day the agile hunter had offered me gold at what worked out at £8 per gram, but the transaction was not really worth pursuing, though this was a fair price. We had no scales and it's difficult to guess a gram of gold by just holding the dust in your hand. The horses swam alongside the canoes and, once across, we took a channel of mud which lasted most of the way to the roadhead leading to Tena. A few miles beyond this point the Rio Jatunyacu joins forces with the Rio Napo. As we neared the end of the path we heard a roaring of an engine and, turning a corner on the trail, there was a large bulldozer thrusting its way towards us through the forest. It was a sobering and frightening sight and drove home to us in no uncertain way the problem of the decimation of the Amazon forest. First barbed wire, then this, I thought!

"It all boils down to too many people in the world," I remarked to Martin as we stopped to take a photo. From the shingle road beyond we got a lift in a pick-up truck to Tena and had our first proper meal in ages.

Tena is a real frontier town, Wild West dirt streets and scruffy Indians, but none looked as scruffy as us. The populace, obviously used to bush-happy Indians visiting Tena's bright lights, had never seen 'white' men like us before. We were indescribably filthy, yet fit and healthy, give or take a few hundred bites and parasitic eggs, which we later excavated from our various bodily crannies.

Our journey to Baños took us back to the risks of bus-riding. It was out of the frying-pan and into the fire! I was fascinated by a red light above the windscreen which came on every time the driver narrowly missed an oncoming vehicle or slid round a corner on loose gravel; it seemed as if it was operated by ESP. Then I realized that it was activated by the brake pedal. Above the light was an effigy of St Christopher. This was obviously an updated and less time-consuming version of the depositing of proceeds of a bus whip-round to a roadside statue of this much-revered saint.

Baños was the end of the trail; we had now wiped the rain forest off our feet. And our remarkable journey had terminated; a hard and dangerous one, for a serious injury on the upper rivers didn't bear thinking about. I wonder if, in retrospect, we did find the location of the old Inca mine? Was the bulk of the treasure now in the Bank of Scotland in Edinburgh? I resolved to beat a path to the illustrious door of that most Royal Bank when I returned.

Back in the old country once more, I made my way across the charming square of St Andrews in Edinburgh. The bank building has an imposing façade set behind heavy gates, and the interior is opulent. There was a surfeit of poker-faced, well-dressed gentlemen in the front hall, looking as if ready to assist a prospective customer with his suitcase of gold bullion. No doubt, despite their immaculate appearance, they can look after themselves and the bank's interest should the need arise. A woman at a desk asked if she could help me.

"I wish to see the Bank Secretary," I explained.

Miami: Joe in jail

The summit crater of Cotopaxi

Left: Market scene near Vilcabamba: Joe in the background

Below: Temporary hold up on the Equadorian highway

Right: A breather in the
Amazon jungle

Below: Trout Camp

At the Great Landslide

Protection from the sun on the first Llanganati trip

The Swamp at Yanacocha

Buachaille Etive Mor, Glencoe, with Crowberry Gully the dark line
running down left of summit

Right: A balloon is laid out on the
summit of Ben Nevis

Below: The cliffs of Ben Nevis

Above: Five Days One Summer: looking down on to one of the high camera platforms in the big crevasse

Left: A climber is landed on the Largo for part of a filming sequence

"I see," she said aloofly. "What exactly is the nature of your business?"

I was at a bit of a loss. I couldn't just say I was here looking for Inca treasure, especially with the flunkies looking on in dignified curiosity. I think they had recognized me and probably wondered what the hell this bearded mountaineer wanted with their bank secretary. A loan for the Glencoe Rescue Team? — the team was always appealing for funds. I cleared my throat. "Well, it's about gold bullion that may have been deposited in this bank about 1803 . . ."

"I see," she said for the second time, as if such a request came half a dozen times before morning coffee. "Well . . . " she hesitated for a moment. "I think you had better see the archivist, Miss Robertson. I'll telephone her." She picked up the telephone. "Yes," she smiled at me after speaking with the archivist. "If you go next door, she will deal with you."

Next door one of the well-dressed ushers was expecting me and he immediately phoned Miss Robertson. She appeared a few minutes later, wearing spectacles and a befitting dress and led me to her office. There I explained my quest and she displayed polite interest, promising to consult the records for me. She was obviously curious when I said that the alleged £460 million deposit was made in 1805.

"That would be a considerable fortune at today's price of gold," she observed primly. "Anyhow, Mr MacInnes, I'll see what I can find out for you . . ."

As promised, she did look up the records and wrote to say that they contained no reference to the treasure. The full text of her letter is reproduced on page 102. But after reading it I was left wondering if Antonio Pastor did find the loot, and, if he did send it on *El Pensamiento*, as indicated, where did it go?

V

Third Time Lucky?

WE BRITS RETURNED to the United Kingdom in February 1980 after
the Llanganati–Rio Napo trip; only Joe Reinhard stayed on alone. He
had thought of going to the States, but as he suffers from perpetual
itchy feet, he was in fact going to Chile to do research on Inca fertility
rites and high-altitude temples as well as working in Peru and
Antarctica.

The Llanganati still preyed on my mind. They seem to affect
everyone in that way. You often hate them when you're there, but
when you leave they seem to call you back. I at least had good reason
to think of them now. The description which Nelson Cerda gave of
that structure, be it smelter or a dinosaur's trough, continued to
fascinate me. There was nothing for it but to return. The *Observer*
newspaper, too, was tentatively still keen, though to date it hadn't
printed a single word on our expedition. That didn't worry us unduly,
provided we could get financial backing.

I contacted the old brigade, as usual by simply picking up the
phone.

"Is that you, Joe?"

"Yes."

"Hamish here. Are you interested in going back up to see that
stonework Cerda found?"

"Yes, when do we go?"

"January."

"Fine."

"Tell Mo, will you, if you see him?"

"Okay."

That was that.

However, cash wasn't quite so easy to obtain now that the
worldwide depression was beginning to cut its teeth, and Martin
couldn't afford to join us. Then I found that Joe Reinhard was in the
wilds of northern Bolivia or some such far-flung place where he
couldn't be contacted easily. We did recruit two others, however;

Mo's wife, Jacky, and Dr Cynthia Williamson, a botanist — both experienced climbers.

To broaden the aims of the expedition, I proposed that we should also try to reach the section of the Verdeyacu river, where I knew the Indians panned gold. This was a short, but difficult journey northwards from the lower section of the Rio Mulatos, over steep escarpments, then a descent into the Yacu gorge. No Europeans had ever been in there. We planned to do that as second fiddle to Cerda's ruins. It was obviously essential to find good porters, after our experience on the last trip with the Parano Indians. I felt, however, that Nelson Cerda could hand-pick strong forest Indians for us.

Brian Warmington was to come along and also a colleague of his, Alan Millar, a climber, who would join forces with us for the second part of the venture.

Back at Baños we made contact with Nelson Cerda via the Amazon radio station and he arranged to meet us at Tena, the last stopping-off town in Amazonia before the green hell.

It was desperately hot when we arrived at Tena; even the flies were taking a siesta in the shade. Tena was just as we had left it, dirt streets, dejected mongrels, stoic mules, Japanese pick-ups and trannies, the latter hammering out music as if they were being sold purely for their volume output.

It was Nelson's son who met us and soon we were picking our way along the jungle trail, past the mission station, to the banks of the Rio Jatunacu. Here we were met by Cerda. He quickly arranged for our party to be ferried across the river in dugouts to La Serena. We set up our tents on some common ground close to Cerda's hut.

With Brian's help we questioned Cerda once again on the structure which he had found. But his tale hadn't changed from the previous February.

"Does he have any theory about it, Brian?" I asked.

"No." Brian translated the answer to my question. "Nelson hasn't any idea what it could have been made for."

"Ask him the exact location again, Brian."

He did so, and replied, "It's about a mile or so above the river in the jungle."

"The river?" I queried.

"That's what he said," Brian affirmed.

"But it's about two miles from the river. Ask him again about the big landslide." I was beginning to have doubts. Had we come all this way for nothing?

"He says, Hamish, that the landslide is further up, beyond the point where you cut off to see the structure."

"I see. This is odd, Brian. Now it doesn't seem to fit with the big landslide. Perhaps he means that smaller landslide that comes into the Rio Mulatos, above the intersection with the Rio Negro?"

"Wait, I'll ask."

"That's it, Hamish," Brian reported. "We were talking about different landslides. He means a small one which goes into the river. He hasn't been up to our big landslide."

"Still," I mused, "it's close enough to the general treasure area; it could have some important significance. I guess we'll see soon enough."

Next morning, in a drizzle, we set off. The Indians proved even better than we had dared hope. Not only could they carry 80 lb. all day in this frustrating terrain, but they did it at twice our pace, and we were all fit. Their uncanny ability to follow the faintest suspicion of a trail which would have foxed a diligent tracker seems second nature to them. A twig broken in a specific way is their rain forest signpost.

We took the precaution of packing two shotguns; rustic instruments made from water tubing, but they must be effective, for the Indians shoot both bear and tapir with them. One can even buy new flint-locks in the Amazonian towns — in this part of the world two packets of cartridges cost more than a gun.

We made a cache of food just above the wide path which we had found towards the end of the journey the previous year. It was here that we proposed to cross the river and attempt to reach the Verdeyacu gorge on the way back. We spent the night here, and the next morning, before the monkeys had rubbed their eyes and had a yawn, we were boulder-hopping up the edge of the Rio Mulatos.

Our journey up that river was well summed up by Brian the previous year, when we had made that rough excursion from the high parama to the Rio Napo. "The only thing that kept me going was the fact that I wouldn't have to repeat the journey." Now, here he was again, hacking his way through some of the worst jungle on earth, climbing along the rocky escarpments as if intent on self-destruction, and wading waist deep in the galloping river.

We were still worried about snakes, especially in the Verdeyacu area, for the Indians said that they were plentiful there. On our way to

Tena from Baños, we called at the mission hospital at Shell Mera to collect some advice and two courses of snake anti-venom. We were amazed to find that one of the mission doctors in Amazonia was successfully treating snake-bite injury by using the HT lead of an outboard motor to administer a shock between the heart and the incision!

It really was soul-destroying going up that river. As before, we camped on clearings by the river edge. At the one where we found Mo incapacitated with his back injury, I had an unfortunate experience. It occurred at dusk during a call of nature. I should perhaps explain that in the jungle one tries to get such necessities behind one (so to speak) before dark. However, this was an emergency. I was keeping a wary eye open for snakes and scorpions, when I chanced to look up at the edge of the rain forest rising abruptly ten feet away. I saw a dusk-blurred furry face peering at me as if through a wreath.

"Good God," I yelled, springing up with complete disregard for dignity and immediately fell flat on my face as I tried to sprint back to camp with my trousers down.

"Hey, Cerda," I shouted, "get your bloody gun. There's a beast in the forest here."

Joe Brown was first out of his tent. Anything to do with jungle wildlife is to him of the utmost importance. He was just in time to see a sapling wave as the uninvited guest beat a hasty retreat to more peaceful surroundings. Cerda scampered off with his gun and a torch like a dark shadow, but returned a short time later without having sighted our crepuscular visitor. He thought that it might have been a bear, or a giant anteater.

We reached the Rio Negro, a tributary of the Mulatos, late one afternoon. Food was getting low and we had to send back all the porters except Cerda in order to conserve supplies. The river was too high to ford; a boiling mass of dirty brown water. The Mulatos, on the other hand, was a greenish blue as it continued upwards to the west. Both colours were, as before, still quite distinctive some way below the confluence.

I was secretly glad of this short pause on the banks of the Negro. It gave me an opportunity to lick my wounds. Though I had been relatively lucky, as I had fared better than some of the others: Brian had a nasty cut from a machete and Cynthia had either fractured or bruised her coccyx. Mo was having eye trouble, which became so bad the next day that he couldn't come with us to the ruins, whilst Jacky had a jippy tummy. Most of us had suffered the onslaught of an ant raid at a lower camp — not actually to our persons, but several of the tents had larger perforations than normal in their mosquito nets and Cerda had some of his clothes eaten. The girls, who had hung their undies to dry on the bushes, also suffered from the ants' attentions. The resulting holes indicated a severe case of 'ants in the pants'!

The following morning the Negro had gone down sufficiently to attempt a crossing and Cerda, displaying that his know-how extended to river crossings, took a safety rope across. We followed and, now in perfect sunshine, tailed him in Indian file at a trot as he took us higher up the Mulatos, then struck inland to the south. After about four hours the angle eased a little and we arrived at some particularly dense under-growth. Brian spoke to Cerda.

"This is it, folks. Not much to see, is there?"

"Sure is overgrown," I agreed. "But there seem to be signs of cultivation." I pointed to where the earth had been dug and nearby trees had been felled.

"He says," Brian translated, "that after we asked him about the ruins last February, he told some of his villagers and they came up to try to make a small farm here."

"Imagine that!" I whistled. "I bet he was trying to establish a claim on the joint. He knew it might be connected with the gold."

"Anyhow," Brian concluded, "Cerda said it was too rough and they couldn't stick it out."

Meanwhile, we had been as interested as kids in a Science Museum. We were peering at holes in the ground, and came to the conclusion that the structure was basically as Cerda had described it. Even his dimensions were fairly accurate. The main part had comprised a trench with stone walls. These were about ten feet high, some nine feet wide and approximately 45 feet long. Leading into it on the uphill side was a channel, some four feet deep and lined with stone, constructed in a V. It extended for at least half a mile uphill and angled towards a subsidiary river. We didn't have time to follow it to its source. At the lower end of the stonework, another channel ran down towards the same river. To each side of the main structure smaller channels led into it. However, the bush was so matted and difficult to clear that we made little impression on it with our machetes and it was too overgrown even to obtain good photographs.

It is quite possible that other buildings exist in the immediate vicinity. In fact, we saw traces of them, but although we had a spade with us, we found it impossible to excavate, owing to the elaborate root systems which lay just below the surface. It would clearly take a major labour force to make any impression on the site.

"Can you make any sense out of this, Joe?" I asked, feeling completely puzzled. I wasn't even sure what to call it, it wasn't a building — a trench?

"Haven't a clue," he returned. He was busy trying to excavate a circular stone-lined hole to the side of the trench and peering in, using his headlamp. "It's a real mystery."

We were all baffled as to what the main structure could have been used for. If, indeed, it was connected with the Inca gold workings. Could it have been a huge gold washing plant? Could the stone-lined pit have been used as a giant sluice box, fed by water from the top channel? It is conceivable that shingle could have been dumped into the structure and the shingle and lighter debris swept down into the lower exit channel by the force of the water. Periodically, the water could be stopped and the gold collected. It is the only solution I could think of.

The location of the gold mine marked on the old Valverde map still appears to me to be where the great landslide occurred, a short way upriver. This landslide is of such magnitude that nothing remains; but these ruins, being so close, could possibly have had some significant part to play in extracting gold.

It was frustrating that, after so much work and a year of 'planning', we had such a short time to spend at the ruins, but that is the way of jungle (and Himalayan) logistics. The area is so inaccessible that you can only take in a limited amount of food. There are not enough porters available to carry more and costs and problems escalate if you try. You end up by having more and more porters to carry more porter food. They won't live on freeze-dried food and they have voracious appetites when food is free, each eating two pounds of sugar a day and corresponding quantities of rice. We had found what we had set out to find; if there was any gold it wasn't readily available, but who knows what may be under that tangle of undergrowth? At any rate, we decided to leave these discoveries for those with more fortitude than we and sadly turned back for camp.

Joe had gone on ahead to try a cast in the river on the way. We said we would see him *en route*. However, as we were threading our way through a great maze of bamboo and other unmentionable obstacles Cerda stopped and pointed. Brian translated.

"Nelson says fisherman went wrong way, he heads east."

Sure enough, Joe had taken a wrong turn, but he didn't admit it until a long time afterwards. In fact he arrived at the river shortly after we did, but he had had a gripping half-hour, fighting through densely packed jungle. He looked even more bedraggled than usual when he caught up with us.

"I thought you were supposed to be getting our dinner?" I quizzed him.

"There's some interesting things in that jungle there." He jerked his telescopic fishing rod in the direction of the green curtain.

"I bet there are," I said dryly.

As we made our way downriver towards the Negro crossing place, we came to a bush on which the Yu Pee berries were ripe. A bear had obviously just vacated the spot, for the sand bore both prints and traces of diarrhoea. The berries were delicious and as I crammed them into my mouth, I said to Brian: "Looking at the effect of these berries on friend Bruno, Brian, these should be called 'Yu-rrhoea' berries."

Next day Mo's vision had improved enough for us to move, so we headed downriver feeling like kids scampering for the playground after the confines of a lengthy session in the classroom. I for one needed to get the grit of the Mulatos and the thorns of the forest out of my skin.

We were not the first to come back from the Llanganati without finding the treasure or Valverde's mine, nor do I suppose we will be the last. As long as the legend persists there will be men — and women — willing to risk entering that alien country to search hopefully. They may in fact find something of greater intrinsic value by finding themselves, for it is only through such hardship, mental or physical, that this is achieved. Perhaps the Llanganati should be retained as an 'International Park of Adventure'. Think of the pleasure it would give to the venturesome unemployed of this world, to be able to go as we did, and pit their wits against nature. But before we go off on other adventures, let us speculate briefly on what I and my friends gleaned from our protracted efforts.

To summarize, it does seem fairly definite that Atahualpa's ransom gold was hidden in the Llanganati and this may, or may not, be based on the same 'mine' as described by Valverde. However, they may be completely separate and the Valverde mine may in fact have been rich gold workings unconnected with the Inca treasure, other than a source of raw material. It's certainly possible that Atahualpa's ransom may have been located by Corregidor Antonio Pastor, who possibly took it to Peru, where in due course it is claimed to have been shipped to the Royal Bank of Scotland in Edinburgh. As the Bank has no trace of this, where did such a vast amount of gold go to? Was it the biggest hijack in history — £460,000,000?

As well as Valverde receiving his wedding present of the gold from his father-in-law, there was the case of Barth Blake who, in April 1887, stumbled on treasure on Cerro Hermoso. Though his companion died, he returned with eighteen pieces of treasure. On his way back on another expedition from England, Blake was lost overboard in a storm. Conveniently, perhaps, for he had his maps and papers with him.

All these are but fragments of the information gleaned over the years, and if any reader goes to the Llanganati to try his or her luck, they may find them useful! But I must add a word of caution: the Llanganati is not a place for a bucket-and-spade visit. It is a serious area, where one can easily come to grief, as history has only too clearly shown. For all but the most determined and experienced, it is better to contemplate such adventures from the sanctity of one's fireside.

> There is something in a treasure that
> fastens upon a man's mind. He will pray and
> blaspheme and still persevere, and will curse the day
> he ever heard of it, and will let his last hour come upon him
> unawares, still believing that he missed it only by a foot. He will
> see it every time he closes his eyes. He will never forget
> it till he is dead and even then . . . There is no
> getting away from a treasure that once
> fastens upon your mind.

> *Nostromo* by Joseph Conrad

Postscript on the Valverde Treasure

BELOW IS THE extract on 'Hidden Treasure' taken from Richard Spruce's notes, written up and discussed by Alfred Russel Wallace, who also knew the area well, in *Notes of a Botanist on the Amazon and Andes*. The map, which appears on pages 96 and 97 of this book, was copied by Spruce from a minute one which he obtained after exhaustive enquiries.

Hidden Treasure

TITLE

GUIDE OR ROUTE WHICH VALVERDE LEFT IN SPAIN, WHERE DEATH OVERTOOK HIM, HAVING GONE FROM THE MOUNTAINS OF LLANGANATI, WHICH HE ENTERED MANY TIMES, AND CARRIED OFF A GREAT QUANTITY OF GOLD; AND THE KING COMMANDED THE CORREGIDORS OF TACUNGA AND AMBATO TO SEARCH FOR THE TREASURE: WHICH ORDER AND GUIDE ARE PRESERVED IN ONE OF THE OFFICES OF TACUNGA

THE GUIDE

"Placed in the town of Pillaro, ask for the farm of Moya, and sleep (the first night) a good distance above it; and ask there for the mountain of Guapa, from whose top, if the day be fine, look to the east, so that thy back be towards the town of Ambato, and from thence thou shalt perceive the three Cerros Llanganati, in the form of a triangle, on whose declivity there is a lake, made by hand, into which the ancients threw the gold they had prepared for the ransom of the Inca when they heard of his death. From the same Cerro Guapa thou mayest see also the forest, and in it a clump of Sangurimas standing out of the said forest, and another clump which they call Flechas (arrows), and these clumps are the principal mark for the which thou shalt aim, leaving them a little on the left hand. Go forward from Guapa in the direction and with the signals indicated, and a good way ahead, having passed some cattle-farms, thou shalt come on a wide morass, over which thou must cross, and coming out on the other side thou shalt see on the left hand a short way off a jucál on a hill-side, through which thou must pass. Having got through the jucál, thou wilt see two small lakes called 'Los Anteojos' (the spectacles), from having between them a point of land like to a nose.

"From this place thou mayest again descry the Cerros Llanganati, the same as thou sawest them from the top of Guapa, and I warn thee to leave the said lakes on the left, and that in front of the point or 'nose' there is a plain, which is the sleeping-place. There thou must leave thy horses, for they can go no farther. Following now on foot in the same direction, thou shalt come on a great black lake, the which leave on thy left hand, and beyond it seek to descend along the hill-side in such a way that thou mayest reach a ravine, down which comes a waterfall: and here thou shalt find a bridge of three poles, or if it do not still exist thou shalt put another in the most convenient place and pass over it. And having gone on a little way in the forest, seek out the hut which served to sleep in or the remains of it. Having passed the night there, go on thy way the following day through the forest in the same direction, till thou reach another deep dry ravine, across which thou must throw a bridge and pass over it slowly and cautiously, for the ravine is very deep; that is, if thou succeed not in finding the pass which exists. Go forward and look for the signs of another sleeping-place, which, I assure thee, thou canst not fail to see in the fragments of pottery and other marks, because the Indians are continually passing along there. Go on thy way, and thou shalt see a mountain which is all of margasitas (pyrites), the which leave on thy left hand, and I warn thee that thou must go round it in this fashion ☞.

"On this side thou wilt find a pajonál (pasture) in a small plain, which having crossed thou wilt come on a cañon between two hills, which is the Way of the Inca. From thence as thou goest along thou shalt see the entrance of the socabón (tunnel), which is in the form of a church porch. Having come through the cañon and gone a good distance beyond, thou wilt perceive a cascade which descends from an offshoot of the Cerro Llanganati and runs into a quaking-bog on the right hand; and without passing the stream in the said bog there is much gold, so that putting in thy hand what thou shalt gather at the bottom is grains of gold. To ascend the mountain, leave the bog and go along to the right, and pass above the cascade, going round the offshoot of the mountain. And if by chance the mouth of the socabón be closed with certain herbs which they call 'Salvaje,' remove them, and thou wilt find the entrance. And on the left-hand side of the mountain thou mayest see the 'Guayra' (for thus the ancients called the furnace where they founded metals), which is nailed with golden nails.★ And to reach the third mountain, if thou canst not pass in front of the socabón, it is the same thing to pass behind it, for the water of the lake falls into it.

"If thou lose thyself in the forest, seek the river, follow it on the right bank; lower down take to the beach, and thou wilt reach the cañon in such sort that, although thou seek to pass it, thou wilt not find where; climb, therefore, the mountain on the right hand, and in this manner thou canst by no means miss thy way."

[Having read this remarkable document, we shall better understand Spruce's account of the various attempts to discover the treasure, the chief routes followed being marked by heavy black lines.]

With this document and the map before us, let us trace the attempts that have been made to reach the gold thrown away by the subjects of Atahualpa as useless when it could no longer be applied to the purpose of ransoming him from the Spaniards.

Pillaro is a somewhat smaller town than Ambato, and stands on higher ground, on the opposite side of the River Patate, at only a few miles' distance, though the journey thither is much lengthened by having to pass the deep quebrada of the Patate, which occupies a full hour. The farm of Moya still exists; and the Cerro de Guapa is clearly visible to east-north-east from where I am writing. The three Llanganatis seen from the top of Guapa are supposed to be the peaks Margasitas, Zunchu, and el Volcan del Topo. The "Sangurimas" in the forest are described to me as trees with white foliage; but I cannot make out whether they be a species of Cecropia or of some allied genus. The "Flechas" are probably the gigantic arrow-cane, *Gynerium saccharoides* (Arvoré de frecha of the Brazilians), whose flower-stalk is the usual material for the Indian's arrows.

The morass (Cienega de Cubillin), the Jucál,† and the lakes called

★[Query — sprinkled with gold. — ED.]

†Júco is the name of a tall, solid-stemmed grass, usually about 20 feet high, of which I have never seen the flower, but I take it to be a species of Gynerium, differing from *G. saccharoides* in the leaves being uniformly disposed on all sides and throughout the length of the stem, whereas in *G. saccharoides* the stem is leafless below and the leaves are distichous and crowded together (almost equitant) near the apex of the stem. The Júco grows exclusively in the temperate and cool region, from 6,000 feet upwards, and is the universal materal for laths and rods in the construction of houses in the Quitonian Andes.

"Anteojos," with the nose of land between them, are all exactly where Valverde places them, as is also the great black lake (Yanacocha) which we must leave on the left hand. Beyond the lake we reach the waterfall (Cascada y Golpe de Limpis Pongo), of which the noise is described to me as beyond all proportion to the smallness of the volume of water. Near the waterfall a cross is set up with the remark underneath, "Muerte del Padre Longo" — this being the point from which the expedition first spoken of regressed in consequence of the Padre's sudden disappearance. Beyond this point the climate begins to be warm; and there are parrots in the forest. The deep dry quebrada (Quebrada honda), which can be passed only at one point — difficult to find, unless by throwing a bridge over it — is exactly where it should be; but beyond the mountain of Margasitas, which is shortly afterwards reached, no one has been able to proceed with certainty. The Derrotero directs it to the left on the left hand; but the explanatory hieroglyph puzzles everybody, as it seems to leave the mountain on the right. Accordingly, nearly all who have attempted to follow the Derrotero have gone to the left of Margasitas, and have failed to find any of the remaining marks signalized by Valverde. The concluding direction to those who lose their way in the forest has also been followed; and truly, after going along the right bank of the Curaray for some distance, a stream running between perpendicular cliffs (Cañada honda y Rivera de los Llanganatis) is reached, which no one has been able to cross; but though from this point the mountain to the right has been climbed, no better success has attended the adventurers.

"Socabón" is the name given in the Andes to any tunnel, natural or artificial, and also to the mouth of a mine. Perhaps the latter is meant by Valverde, though he does not direct us to enter it. The "Salvaje" which might have grown over and concealed the entrance of the Socabón is *Tillandsia usneoides*, which frequently covers trees and rocks with a beard 30 or 40 feet long.

Comparing the map with the Derrotero, I should conclude the cañon, "which is the Way of the Inca," to be the upper part of the Rivera de los Llanganatis. This cañon can hardly be artificial, like the hollow way I have seen running down through the hills and woods on the western side of the Cordillera, from the great road of Azuáy, nearly to the river Yaguachi. "Guayra," said by Valverde to be the ancient name for a smelting-furnace, is nowadays applied only to the wind. The concluding clause of this sentence, "que son tachoneados

de oro," is considered by all competent persons to be a mistake for
"que es tachoneado de oro."

If Margasitas be considered the first mountain of the three to which
Valverde refers, then the Tembladál or Bog, out of which Valverde
extracted his wealth, the Socabón and the Guayra are in the second
mountain, and the lake wherein the ancients threw their gold in the
third.

Difference of opinion among the gold-searchers as to the route to
be pursued from Margasitas would appear also to have produced
quarrels, for we find a steep hill east of that mountain, and separated
from it by Mosquito Narrows (Chushpi Pongo), called by Guzman
"El Peñon de las Discordias."

If we retrace our steps from Margasitas till we reach the western
margin of Yana-cocha, we find another track branching off to
northward, crossing the river Zapalá at a point marked Salto de
Cobos, and then following the northern shore of the lake. Then
follow two steep ascents, called respectively "La Escalera" and "La
Subida de Ripalda," and the track ends suddenly at the river coming
from the Inca's Fountain (La Pila del Inca), with the remark,
"Sublevacion de los Indios — Salto de Guzman," giving us to
understand that the exploring party had barely crossed the river when
the Indians rose against them, and that Guzman himself repassed the
river at a bound. These were probably Indians taken from the towns to
carry loads and work the mines; they can hardly have been of the
nation of the Curarayes, who inhabited the river somewhat lower
down.

A little north and east of the Anteojos there is another route running
a little farther northward and passing through the great morass of
Illubamba, at the base of Los Mulatos, where we find marked El
Atolladero (the Bog) de Guzman, probably because he had slipped up
to the neck in it. Beyond this the track continues north-east, and after
passing the same stream as in the former route, but nearer to its source
in the Inca's Fountain, there is a tambo called San Nicolas, and a cross
erected near it marks the place where one of the miners met his death
(Muerte de Romero). Another larger cross (La Cruz de Romero) is
erected farther on at the top of a basaltic mountain called El Sotillo. At
this point the track enters the Cordillera de las Margasitas, and on
reaching a little to the east of the meridian of Zunchu-urcu, there is a
tambo with a chapel, to which is appended the remark, "Destacamento de Ripalda y retirada per Orden Superior." Beyond the fact

thus indicated, that one Ripalda had been stationed there in command of a detachment of troops, and had afterwards retired at the order of his superiors, I can give no information.

There are many mines about this station, especially those of Romero just to the north, those of Viteri to the east, and several mines of copper and silver which are not assigned to any particular owner. Not far to the east of the Destacamento is another tambo, with a cross, where I find written, "Discordia y Consonancia con Guzman," showing that at this place Guzman's fellow-miners quarrelled with him and were afterwards reconciled. East-north-east from this, and at the same distance from it as the Destacamento, is the last tambo on this route, called El Sumadal, on the banks of a lake, near the Rio de las Flechas. Beyond that river, and north of the Curaray, are the river and forests of Gancaya.

Another track, running more to the north than any of the foregoing, sets out from the village of San Miguel, and passes between Cotopaxi and Los Mulatos. Several tambos or huts for resting in are marked on the route, which ends abruptly near the Minas de Pinel (north-east from Los Mulatos), with the following remark by the author — "Conspiracion contra Conrado y su accelerado regreso," so that Conrado ran away to escape from a conspiracy formed against him, but who he was, or who were his treacherous companions, it would now perhaps be impossible to ascertain.

Along these tracks travelled those who searched for mines of silver and other metals, and also for the gold thrown away by the subjects of the Inca. That the last was their principal object is rendered obvious by the carefulness with which every lake has been sounded that was at all likely to contain the supposed deposit.★

The mines of Llanganati, after having been neglected for half a century, are now being sought out again with the intention of working them; but there is no single person at the present day able to employ the labour and capital required for successfully working a silver mine, and mutual confidence is at so low an ebb in this country that companies never hold together long. Besides this, the gold of the Incas never ceases to haunt people's memories; and at this moment I am informed that a party of explorers who started from Tacunga imagine they have found the identical Green Lake of Llanganati, and are preparing to drain it dry. If we admit the truth of the tradition that the ancients smelted gold in Llanganati, it is equally certain that they extracted the precious metal in the immediate neighbourhood; and if the Socabón of Valverde cannot

★The soundings of the lakes are in Spanish varas, each near 33 English inches.

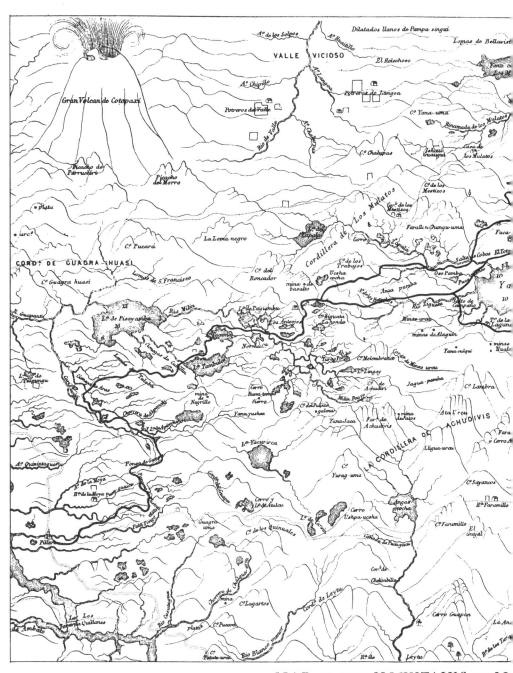

MAP OF THE MOUNTAINS OF LL

by Do

To illustrate a

(Jour

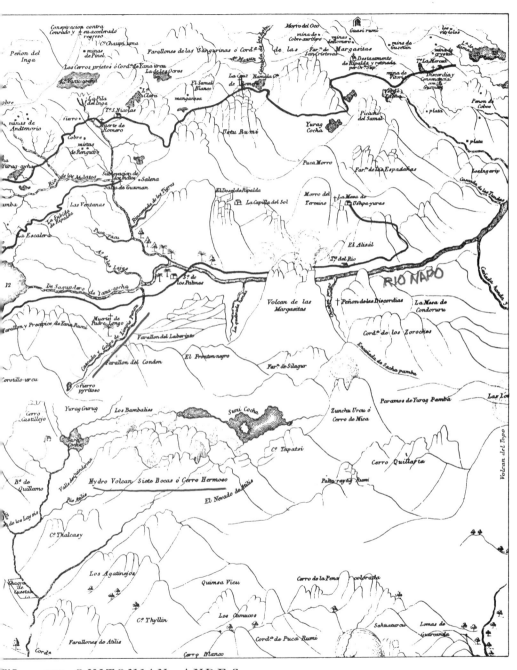

TI, IN THE QUITONIAN ANDES

...uzman.

...rd Spruce Esq.ᵣₑ

...iety)

at this day be discovered, it is known to every one that gold exists at a short distance, and possibly in considerable quantity, if the Ecuadoreans would only take the trouble to search for it and not leave that task to the wild Indians, who are content if, by scooping up the gravel with their hands, they can get together enough gold to fill the quill which the white man has given them as the measure of the value of the axes and lance-heads he has supplied to them on trust.

The gold region of Canelos begins on the extreme east of the map of Guzman, in streams rising in the roots of Llanganati and flowing to the Pastasa and Curaray,* the principal of which are the Bombonasa and Villano. These rivers and their smaller tributaries have the upper part of their course in deep ravines, furrowed in soft alluvial sandstone rock, wherein blocks and pebbles of quartz are interspersed, or interposed in distinct layers. Towards their source they are obstructed by large masses of quartz and other rocks; but as we descend the stones grow fewer, smaller, and more rounded, until towards the mouth of the Bombonasa, and thence throughout the Pastasa, not a single stone of the smallest size is to be found. The beaches of the Pastasa consist almost entirely of powdered pumice brought down from the volcano Sangáy by the river Palora. When I ascended the Bombonasa in the company of two Spaniards who had had some experience in mining, we washed for gold in the mouth of most of the rivulets that had a gravelly bottom, as also on some beaches of the river itself, and never failed to extract a few fragments of that metal. All these streams are liable to sudden and violent floods. I once saw the Bombonasa at Pucayacu, where it is not more than 40 yards wide, rise 18 feet in six hours. Every such flood brings down large masses of loose cliff, and when it subsides (which it generally does in a few hours) the Indians find a considerable quantity of gold deposited in the bed of the stream.

The gold of Canelos consists almost solely of small particles (called "chispas," sparks), but as the Indians never dig down to the base of the wet gravel, through which the larger fragments of gold necessarily percolate by their weight, it is not to be wondered at that they rarely encounter any such. Two attempts have been made, by parties of Frenchmen, to work the gold-washings of Canelos systematically. One of them failed in consequence of a quarrel which broke out among the miners themselves and resulted in the death of one of them. In the other, the river (the Lliquino) rose suddenly on them by night and carried off their canoes (in which a quantity of roughly-washed gold

*The name Curaray itself may be derived from "curi," gold.

was heaped up), besides the Long Tom and all their other implements.

Critical Note by the Editor

The preceding account of the various routes of the gold-seekers among the Llanganati Mountains leads to the conclusion that only the earliest — that led by the Corregidor of Tacunga and the friar Padre Longo — made any serious attempt to follow the explicit directions of the "Guide," since the others departed from it so early in the journey as the great black lake "Yana Cocha," going to the left instead of to the right of it. No doubt they were either deceived by Indian guides who assured them that they knew an easier way, or went in search of rich mines rather than of buried treasure. The first party, however, and those who afterwards followed it, kept to the route, as clearly described, to the sleeping-place beyond the deep ravine where Padre Longo was lost; but beyond this point they went wrong by crossing the river, and thus leaving the district of the three volcanoes, which twice at the beginning of the "Guide" are indicated as the locality of the treasure.

Although no route to these mountains is marked on the map, Spruce tells us that other parties did take the proper course, and found the "deep dry ravine" (marked on the map as "Quebrada honda"), and after it the mountain of Margasitas; but here they were all puzzled by the "Guide" directing them to leave the mountain on their left while the hieroglyph seems to leave it on the right, and following this latter instruction they have failed afterwards to find any of the other marks given by Valverde in his "Guide." Spruce himself suggests that the upper part of the Rivera de los Llanganatis (which is outside the portion of the map here given) is the "way of the Inca" referred to in the "Guide." But this is going quite beyond the area of the three mountains, so clearly stated as the objective of the "Guide."

It seems to me, however, that there is really no contradiction between the "Guide" and the map, and that the route so clearly pointed out in the former has not yet been thoroughly explored to its termination, as I will now endeavour to show. After crossing the deep dry ravine ("Quebrada honda" of the map), we are directed to "go forward and look for the signs of another sleeping-place." Then, the next day — "Go on thy way, and thou shalt see a mountain which is all of margasitas, the which leave on thy left hand." But looking at the map we shall see that the mountain will now be on the right hand, supposing we have gone on in the same direction as before. crossing the deep ravine. The next words, however, explain this apparent contradiction: they are — "and I warn thee that thou must go round it in this fashion," with the explanatory hieroglyph, which, if we take the circle to be the mountain and the right-hand termination of the curve the point already reached, merely implies that you are to turn back and ascend the mountain in a winding course till you reach the middle of the south side of it. So far you have been going through forest, but now you are told — "On this side thou wilt find a pajonál (pasture) in a small plain" (showing that you have reached a considerable height), "which having crossed thou wilt come on a cañon between two hills, which is the way of the Inca." This cañon is clearly the upper part of the "Chushpi pongo," while the "Encañado de Sacha pamba" is almost certainly the beginning of the "way of the Inca." The explorers will now have reached the area bounded by the three volcanoes of the "Guide" — the Margasitas will be behind them, Zunchuurcu on his right, and the great volcano Topo in front, and it is from this point only that they will bc in a position to look out for the remaining marks of the "Route" — the socabón or tunnel "in the form of a church porch," and evidently still far above them, the cascade and the quaking-bog, passing to the right of which is the way to "ascend the mountain," going "above the cascade" and "round the offshoot of the mountain" to reach the socabón. Then you will be able to find the Guayra (or furnace), and to reach the "third mountain," which must be the Topo, you are to pass the socabón "either in front or behind it, for the water of the lake falls into it." This evidently means the lake mentioned in the first sentence of the "Guide" as being the place where the gold prepared for the ransom of the Inca was hidden. The last sentence of the "Guide" refers to what must be done if you miss the turning shown by the hieroglyph, in which case you have to follow the river-bank till you come to the cañon (on the map marked "Chushpi pongo"), up the right-hand side of which you must climb the mountain, "and in this manner thou canst by no means miss thy way"; which the map clearly shows, since it leads up to the "Encañado," which is shown by the other and more easy route to be the "way of the Inca."

I submit, therefore, that the "Guide" is equally minute and definite in its descriptions throughout, that it agrees everywhere with Guzman's map, and that, as it is admitted to be accurate in every detail for more than three-fourths of the whole distance, there is every probability that the last portion is equally accurate. It will, of course, be objected that, if so, why did not Guzman himself, who made the map, also complete the exploration of the route and make the discovery? That, of course, we cannot tell; but many reasons may be suggested as highly probable. Any such exploration of a completely uninhabited region must be very costly, and is always liable to fail near the end from lack of food, or from the desertion of the Indian porters when there was doubt about the route. Guzman had evidently been diverted from the search by what seemed the superior promise of silver and gold mines, from which he may have hoped to obtain wealth enough to carry out the other expedition with success. This failing, he apparently returned home, and may have been endeavouring to obtain recruits and funds for a new effort when his accidental death occurred.

It is to be noted that beyond the point where the hieroglyph puzzled all the early explorers there is a complete absence of detail in Guzman's map, which contains nothing that might not have been derived from observations made from the heights north of the river, and from information given by wandering Indians.

It is also to be noted that only four sleeping-places are mentioned in the "Guide," so that the whole journey occupied five days. The last of the four sleeping-places is before reaching the spot where the path turns back round the Margasitas Mountain, so that the whole distance from this place to the "lake made by hand" must be less than twenty miles, a distance which would take us to the nearer slopes of the great Topo Mountain. In this part of the route the marks given in the "Guide" are so many and so well-defined that it cannot be difficult to follow them, especially as the path indicated seems to be mostly above the forest-region.

For the various reasons now adduced, I am convinced that the "Route" of Valverde is a genuine and thoroughly trustworthy document, and that by closely following the directions therein given, it may still be possible for an explorer of means and energy, with the assistance of the local authorities, to solve the interesting problem of the Treasure of the Incas. The total distance of the route, following all its sinuosities, cannot exceed ninety or a hundred miles at most, fully three-fourths of which must be quite easy to follow, while the remainder is very clearly described. Two weeks would therefore suffice for the whole expedition.

I have written this in the hope that some one who speaks Spanish fluently, has had some experience of the country, and is possessed of the necessary means, may be induced to undertake this very interesting and even romantic piece of adventurous travel. To such a person it need be but a few months' holiday.

★

Letter to the author from the archivist of the Royal Bank of Scotland regarding the alleged £460 million gold deposit made in 1805.

Royal Bank of Scotland Limited

H. MacInnes, Esq.,
Glencoe,
Argyll.

Dear Mr MacInnes,

Estates of Antonio Pastor y Marin de Segura and his wife,
Narcisa Martinez de Tajada y Oraye

I refer to your enquiry regarding the 'Valverde treasure' — which I assume is a reference to the above. We had a great many enquiries regarding this 'treasure' during the years 1965/1968, principally from South America (largely from Peru). The details as known to us are:—

Corregidor Antonio Pastor y Marin de Segura, Marquesa of Llosas, was apparently a highly-placed official in the service of the King of Spain in South America before the independence of the present Latin-American states and lived there in various places from 1794 until his death in 1804.

He is said to have shipped 'treasure' (precious stones, silver, gold, etc.) to the value of £460,000,000 to Scotland in 1803, in the *El Pensamiento* under the joint captaincy of John Doigg and John Fanning; who are said to have deposited it in 1803 through the agency of Sir Francis Mollison (or Mollinson), described as a Banker, in The Royal Bank of Scotland (Banco Real de Escocia) in Edinburgh.

It is also said that with the 'treasure' went instructions that it should be held for the fifth generation of Antonio Pastor and his wife, Narcisa Martinez — he apparently did not get on well with his son.

The *El Pensamiento* is likely to have been a captured Spanish ship sailing under the British flag. We confirmed at the time that there was no record of the ship in Lloyd's Register of Shipping.

We made a thorough search of our records here at the time and also had a search made at our Buchanan Street Glasgow Branch, the only Branch of The Royal Bank of Scotland in existence at the period in question, but without throwing any light on the story. There were numerous references to it in the South American and Spanish Press during 1965/1968.

I hope that this information is of assistance to you, but if I can be of any further help, then please let me know.

Yours sincerely,
C. H. Robertson (Miss)
Librarian and Archivist.

★

With the assistance of the publishers, I had various searches conducted in Britain and in Spain to see if the *El Pensamiento*, the ship which reputedly carried the treasure from Peru to Scotland, could be traced. Also, an attempt was made to locate the Royal Warrant issued by the King of Spain, which authorized the original expedition to procure the Valverde treasure.

Regarding the *El Pensamiento* investigation, I'm indebted to the assistance given by Mr Fred Walker and his staff at the National Maritime Museum, London, Miss Anne Escott and colleagues of the Mitchell Library, Glasgow, and other professional researchers in Scotland.

This enquiry revealed that the *El Pensamiento* is not recorded in *Lloyd's Register*, *Lloyd's List*, the *Shipping News*, *Glasgow Courier* or the *Edinburgh Courant* during the years 1802–04.

This points (but not conclusively) to the fact that no ship under the name *El Pensamiento* docked at a Scottish port during this period. However, the Museum Historian of the National Maritime Museum suggests that around 1803 it is possible that ships sailing from Peru may have given false information on their clearance papers and manifests. If treasure was being shipped, it must have been in another vessel or, if in the *El Pensamiento*, sailing under another name. However, officials of the Royal Bank of Scotland state that, according to their records, no consignment of treasure was ever received by them.

Another series of records (E501–515), which covers the period in question, should be housed in the Scottish Records Office in Edinburgh. However, a researcher has established that these are missing for the years October 1796 to October 1805.

Considering the above information, it's my impression that the Valverde treasure is most likely still languishing in the Llanganati of Ecuador, less the percentages of Valverde and Blake.

Now let us turn our attention to the Royal Warrant. With the help of experts in Hispanic studies, I have discovered that this document could possibly still be in archives in Spain, in one of several trunks containing uncatalogued eighteenth-century material. The careful investigation of this will be a long and tedious task and I have requested an Official Search for these documents. Should this be granted by the Spanish authorities, it could shed new light on this old and fascinating mystery, one of the most intriguing of all treasure stories. If, as some experts believe, the contents of the original Cedula Real (Royal Warrant) were altered to confuse later aspirant adventurers, a detailed study of the original (should we find it) could provide fresh and exciting clues to this puzzle, a puzzle as exasperating as the terrain of the Llanganati.

I am most grateful to Dr Ivy McClelland, Dr Ann L. MacKenzie, Professor Dorothy Sherman Severin, Professor Juan Mercader Riba, Rosario Parria and Dudley Knowles. These scholars helped me in the Royal Warrant search and suggested lines of investigation known only to experts in the labyrinth of historical fact-finding.

BACK HOME

MY HOME IS in Glencoe in Scotland. Many people regard this historical glen as a wild place with brooding mountains; to me it is a glen of infinite variety, be it in blizzard, sunshine or torrential rain. It can be dark; yes, with clouds hugging the rocks, especially in late autumn, but it comes to life in snow. The peaks appear new, refreshed and welcoming. The glen rests under the protective wing of the National Trust for Scotland, who are probably the ideal landowners, being neither too restrictive nor too lax in their custody of this unique place.

Glencoe has its own particular history, and quite a bloody one. Each time I use the back door of my home, I'm reminded of man's inhumanity to man, for five MacDonalds, one of them a child, were murdered here in the infamous Massacre of Glencoe. On a cold February in 1692, Campbell troops, billeted with the local MacDonalds, and under a pretence of friendship, turned on their hosts during the night, cutting down men, women and children. This heinous crime, 'Murder under Trust', was the most deplored in the old annals of Scottish Law, and even to this day Campbells are still distrusted in the Glen.

Today, lives are lost in the glen in a different fashion — in accidents. Now the victims are climbers and we can't blame the Campbells! These accidents, tragic as they are, are often associated with courage and compassion, and are a way of life to the local mountain rescue team. Rescue work is something which the team enjoys. That's not to say that we relish the prospect of someone being injured or killed. We do enjoy the company (the crack) of fellow climbers and the bond which is forged by a band of fit men working together. The fact that their skills can be focused to help someone makes it all worthwhile, and as most of the team are climbers, there is always the thought, "But for the Grace of God . . ."

I suppose it is natural, when dealing with the high stakes of human life, that certain aspects of the operation are focused and retained in the memory with crystal clarity. Often, though, it is only when we are struggling up the slope or perhaps awaiting the arrival of the helicopter on some forlorn hillside, that recollections of a previous rescue come tumbling back.

Interestingly, too, it is usually the humorous aspects that are conjured up, not the gore and mangled limbs. It may be how a team member arrived with two left boots or, for a recent example, when we went up to rescue a man who had fallen in Clachaig Gully. We got up to the climber and put him on one of my stretchers. He was lifted out by helicopter and on the way back from hospital the chopper called in at my place. The winchman handed me a note scribbled on an RAF memo pad, the type which aircrew have strapped to their knees. It was from the casualty. He wrote that he was sorry to have met me in such unfortunate circumstances, but had actually come to Glencoe to buy one of my stretchers.

"At least," I said to the winchman with a laugh, "he got a real demonstration!" As I never did receive an order from him, I can only conclude that he has now given up climbing or the MacInnes Mk. V stretcher wasn't to his satisfaction.

Perhaps I should say something here about how the Rescue Team was founded. The Glencoe Mountain Rescue Team, to give its full title, was started in 1959. This 'official' move was primarily to try to acquire equipment, for in those days even the village bobby used to accompany us, dressed in his regulation uniform and a pair of wellingtons.

Of course, rescues had taken place in the Glen before. Walter and Willie Elliot had followed in their father's footsteps on such mercy missions since they were boys. Both knew the hills as only shepherds can — the easy way to gather sheep from the cliff faces is usually the ideal way to evacuate a casualty. Willie, now the National Trust Ranger, as well as gamekeeper, has gained some weight since those early years with his father, but he still operates our base truck or Landrover with unabated efficiency. He knows the complicated terrain of Glencoe as well as his own living-room or the local hostelries. His brother Walter is quiet and slim, a gentle person of interminable patience. Here is a man who cares for his flock. Competent on the hill, he has precise judgement of his own capabilities and of danger, which could be the envy of a mountaineer. Both Walter and Willie are unmarried and live with their sister, Doris, in the tiny snug cottage at Achnabeith which seems to purr by the side of Loch Achtriochtan in the very heart of the Glen. The only other active founder-member of the team is Denis Barclay. In those early days, Denis worked at the ski tow, but now is employed by the North of Scotland Hydro-Electric Board. Denis's passions in life, which he shares with his wife, Iris, are skiing and sunshine. At every available opportunity they steal off to some clime which offers one or both of these ingredients, depending upon the season. Their knowledge of holiday resorts is vast and rivals that of a competent travel agent. Denis, though he makes no pretence at being a mountaineer, is no slouch on the hills, and more than once has gone with alacrity where both angels and mountaineers with less bottle have feared to tread.

It doesn't take long for a new member of the Team to become 'blooded', and used to accidents. Sometimes this happens in a gentle way, with perhaps nothing more gruesome than a broken leg or an exposure case. Occasionally, however, they are baptized at the deep end. For example, Peter Weir's first rescue experience was not for the faint of heart. Peter is one of those salt-of-the-earth men, generous and always willing to carry the stretcher. Being big and strong, he is a tremendous asset to the team even though he's not a climber. The rescue was on Buachaille Etive Mor in winter time. I had a call to say that a climber had fallen on the Buachaille, not far above the access path to the mountain, where it crosses a subsidiary gully. I wasn't sure exactly where the accident had occurred, but before leaving the house, I asked for a helicopter from RAF Leuchars in Fife. From here, a flight of Wessex helicopters serves us and many other rescue organizations throughout Scotland. The Wessex is smaller than the Sea King, which is the other helicopter in general rescue use with the RAF. The Leuchars pilots and crew have, over the years, achieved a very high standard of efficiency, sometimes having to fly in appalling weather conditions which only a few years ago would have been impossible, and which I often think still are.

I had gone up ahead of the others as I was closer to the mountain and started to search the face above the Buachaille path. Presently, I saw a couple of figures in an ice-filled gully. It wasn't a known route, but a runnel of ice. I assumed that they had climbed it as a practice route, for the weather that day wasn't good higher up. Later I was to learn that they had confused this ice runnel with the prodigious Great Gully, further up the mountain. It was comparable to mistaking an alley for the M1.

I climbed up to where they were, only putting crampons on for the last bit. There were two climbers there, members of the injured man's party, one beside the casualty and the other a short way below. The injured man was slumped back against the ice on a small ledge and there wasn't a great deal of room. He was unroped.

"How is he?" I asked the top man when I got alongside.

"I think he's dead," he replied despondently, "but I'm not sure."

"Here, let me have a look."

Below, I caught a glimpse of some of the team coming up the side of the ice runnel as I had done.

"You go down to your mate now," I told the climber. "We'll take care of things. Can you manage okay?"

"Yes, I think so."

I quickly examined the fallen climber and thought I could detect a very faint pulse. I tried to resuscitate him using my airway, but he was in a hell of an awkward position and I couldn't get him into a more convenient posture on my own. Taking up my walkie-talkie mike, I called Willie Elliot; I had seen him arriving with the base truck.

"Willie, can you give someone the oxygen gear? This chap needs it urgently."

In my heart of hearts, I knew full well that by the time it arrived, this poor lad would be past anything but divine assistance, but one had to go through the motions. Peter Weir was now just below, moving like a steam engine. Behind him were Larry Taylor, David Gunn and Bob Hamilton. Bob, who at one time studied ballet, has now turned his talents to prawn fishing. Larry and David were both woodsmen, Larry an ardent fell runner who had been running up several mountains a week in training, so, needless to say, both were fit as fleas.

"Give me a hand, Peter. I'm afraid I don't have much hope for this guy, though."

"I'll be there in a minute, Hamish. The others are just behind."

One of the friends of the casualty shouted up, "I can hear a helicopter." Sure enough, that familiar throb could be heard getting louder each second as the large machine beat its way across the Moor of Rannoch. The 'wuff wuff' was quite distinct now. I got on the radio.

"Willie, get the chopper on the radio and ask them to lower oxygen equipment down to us. We'll see if we can revive this bloke, he's in a critical condition. I'm out of radio contact here as we're between the rock walls of this wee gully."

"I'll do that, Hamish."

In a few minutes the helicopter came into view. Willie had made contact and had given them our position. In the meantime, Peter was helping me manhandle the unconscious man into a more comfortable position. He still gave no response to the airway or cardiac massage.

The helicopter came nearer and we could see the winchman peering out from the cabin. The machine manoeuvred into position some 200 feet above and, after what seemed an interminable delay (but was only two minutes), the winchman swung into view and slowly spiralled down. It was obviously going to be difficult for him to land nearby, for the rocks were coated in water ice, but Bob managed to field him on to a small rock ledge just below, close to where the two survivors were. The winchman had the oxygen and Bob quickly climbed up with it. In a few seconds I had the mask over the injured man's face. He had apparently lost his helmet in the fall, but he still had a balaclava pulled down, leaving only part of his face exposed. The mask hissed as I turned on the cylinder. I gave maximum pressure, about 20 litres per minute, and started rhythmically squeezing the resuscitator bulb. Peter was holding him in position. To my consternation, I saw his balaclava pulsate like a concertina, in unison with the resuscitator pressure. I eased the edge of his balaclava up above his forehead and saw that a scalp wound had formed a huge trapdoor through which the oxygen leaked. Later we deduced that the ferrule end of his ice axe must have entered his mouth when he fell and penetrated the roof of his mouth to exit through the temple, probably knocking off his climbing helmet. The axe must have been subsequently dislodged in the continuation of the fall. "Not a very pleasant sight, Peter, but if you've got to go, you've got to go."

Peter makes a point of drawing attention to the fact that he isn't a 'climber', yet he seems to manage remarkably well in difficult situations. On one of the Clachaig Gully rescues, the injured climber lay at the base of the last pitch of this rather damp and vegetative climb. Peter abseiled down to assist some of the team with the evacuation. Meanwhile, in the Clachaig Inn below, some of the team were standing by in the bar in case they were required, partaking of the good ale. Ronnie Rodger, the son-in-law of the hotelier Mike Gardiner, had propped his walkie-talkie on the bar and the assembled patrons and the team could overhear the talk-back between those in the Gully and myself and Willie in the Landrover at the base of the climb.

Ronnie, who has a passion for radio equipment, cut in and asked Peter, "How do you enjoy climbing in the dark?"

"I'll tell you something." Peter's voice came over loud and clear and very emphatic. "I'm relying on Paul Moores and Davy Gunn to get me out of this hole, and I'll need a few rolls of toilet paper, I can tell you!"

The vast majority of letters which I get are from distressed relatives of accident victims, and these make one acutely aware of the sorrow caused by many of these accidents. But I must admit that other letters can be very amusing, providing much-needed comic relief in this line of duty.

In the five-year span which this book covers, we had many interesting rescues; probably about 40 would be worth recording in detail. The following story of rescue work in the Glen was staged before the great auditorium of the Rannoch Moor, as it took place on Buachaille Etive Mor. It was, alas, a tragedy, but then again rescues don't often have happy endings.

The storms which we experienced whilst attempting the Ben Nevis live TV broadcast (described in the book's final section) continued into the spring of 1982. Only a few interludes were suitable for winter climbing, but the late snow lingered in the higher gullies well into April. Indeed, the climbing conditions were good and my girlfriend and I climbed Crowberry Gully in early April, which was most unusual. Hard snow persisted, with only one pitch baring its rock for the coming summer. There were, however, great steps leading up from the junction of Crowberry Gully and Easy Gully. Later I discovered that one of our rescue team members had taken a large party of soldiers up the gully in a training exercise two days before our climb — hence the Hannibal-like stairway.

Perhaps I'd better give a brief explanation of this part of the mountain. Buachaille Etive Mor stands alone at the very western extremity of the Moor of Rannoch, acting as a corner-post between Glencoe and Glen Etive. To those who give more than a passing glance to the mountain as they speed along the A82, a thumb of rock can be seen sticking up close to the summit facing the Moor. This is Crowberry Tower. To its right, Crowberry Ridge runs down the face for all of 700 feet. And to the right of that is Crowberry Gully, a deep defile 1,000 feet long, part of the mountain's drainage system. In winter it fills with snow and ice and offers one of the finest winter climbs in Scotland. A magnificent classic route, it was first climbed in 1936 and then classified as severe. It is a three-star outing boasting panoramic views; there are usually a dozen or so pitches offering varying degrees of difficulty, depending on the build-up of ice. It branches at the top, the left-hand route going up to the small breche behind Crowberry Tower with the main gully sweeping majestically between rock walls to emerge within snowball-throwing distance of the summit of the mountain.

I was working on the last of my three West Highland Walks books when I got the rescue call from the police. There wasn't much to go on — our only report was that someone had fallen down Crowberry Gully and that he was suffering from head injuries. After requesting a helicopter, I immediately contacted Doris Elliot for a team call-out. Helicopters have become regular work horses of modern-day mountain rescue, but we have a policy in the Glencoe Team that, irrespective of a helicopter *en route*, an advance party sets off on foot. It's just possible that the chopper may be re-routed to somewhere else such as a plane crash, which would have priority, or some other military requirement. Helicopters are used for rescue work in Britain upon request to the military authorities. Usually these are from RAF stations, but both Naval and Marine helicopters are also used, all free of charge. They are of tremendous advantage to the rescue teams and these operations are excellent training for pilots and crew.

When I turned into the car park at Jacksonville, there was only one other car there, and it wasn't one of ours. I drew alongside it, peering up at the mountains through the windscreen to see if I could make out any figures on the mountain. I heard a voice.

"It's all right, Hamish." There was an accompanying laugh. "I've got the rescue operation under control."

It was Dudley Knowles, a lecturer at Glasgow University. He had been a member of the rescue team some eight years previously. His identical twin brother, Dave, was also one of my instructors when I had my climbing school in the Glen. Dave had been killed filming in Switzerland.

"Hi! Great to see you, Dudley," I shouted as I climbed out of the car. "Heard anything of this accident?"

"No. I never knew there was one until you hurtled into this lay-by. I'm just taking my dad on a tour of the old haunts."

"Well, I'd better be off," I said. "I gather that this chap's in a bad way. I don't know if I'll get up before the helicopter arrives or not." Just as I set off, the police Range Rover skidded to a halt. The other team members were on their way.

It was a fine day and I climbed the steep path by the Waterslab on the

lower slopes of the mountain. I could see across the expanse of the
Moor of Rannoch to distant Shiehallion, still splattered in white like a
ptarmigan in its spring apparel. Glancing back at the lay-by, I saw a
number of cars parked and our rescue truck just turning in. I knew
Willie Elliot was driving. Ronnie Rodger's voice crackled from my
walkie-talkie:

"Hello, Hamish, Ronnie here. What else do we need to take up?"

"The stretcher, Ronnie, and the first-aid rucksack which has the
casualty bag in it." I didn't know the situation then. "Oh, better take a
rope, too."

"Okay. Peter Weir, Ian Nicholson and myself are just setting off."

"I'll keep in touch, Ronnie."

Ronnie Rodger has been rescuing for thirteen years. He is one of the
few 'true' local men on the team. His great-great-grandfather, a local
shepherd named Neill Marquis, was the first man to climb up to
Ossian's Cave on Aonach Dubh in Glencoe in 1882. The cave is a
rather hazardous place to reach, so good climbing instinct must run in
Ronnie's family.

As I emerged over an easy rock section, I saw the figure of a man
above and, as I approached, I could see another, the second lying
prostrate across a rocky rib. This second figure seemed to be perched
rather precariously, half draped over the edge of the rocks like a
discarded rag doll. He didn't look in the best of health.

"How is he?" I shouted to the climber who stood next to the fallen
man.

"I think he's dead," came the response. "He seemed to stop
breathing a short time ago."

"Bloody hell," I muttered. In the distance I could hear the throb of
the helicopter.

I quickly climbed up to where the casualty lay and found, to my
surprise, that he was a youth. 'He can't be more than eighteen,' I
thought. As the climber beside him had said, there seemed no sign of
life, but I still decided to try mouth-to-mouth resuscitation and
whipped out an airway I had in my rucksack, which I keep there with
my first-aid. I always carry an airway, which connects the rescuer's
mouth to that of the victim. I use one whenever possible because, I
must admit, one of the most nauseating experiences I know is the
lingering taste from a corpse which you have tried to revive with the
kiss of life. If the casualty recovers, it doesn't seem to matter, but if he
or she dies, that flavour of death persists and seems to haunt one for

months. For that reason, I always use the 'Brooke's Airway', as it incorporates a one-way valve which prevents vomit and mucus from passing through the tube.

I quickly forced the tube between his tightly-clamped jaws and began the steady, hopeful, let's-pray-I'm-not-too-late breathing, familiar to all who deal with such emergencies.

"Were you with him?" I asked the climber, between breaths.

"No, I was on my way down Curved Ridge when I heard the shouts for help. I then saw this chap here. It was his friends who were shouting for help. They're up in Crowberry Gully. My pal went down to give the alarm. There he is below," he pointed, "he's coming back up."

Just behind this man were Ian Nicholson, Ronnie Rodger and Peter Weir, who was carrying the stretcher. They had come up like express trains. I grabbed my walkie-talkie mike and called the chopper direct.

"Hello, Helicopter Three-Four, this is the Glencoe hill party, Hamish speaking. Do you read me? Over."

"Helicopter Three-Four. Hello, Glencoe hill party. Reading you loud and clear."

"Can you fly directly up to me at the casualty location? Directly under Crowberry Ridge, where the snow of Crowberry Gully stops. I have a fluorescent wind marker. Request oxygen equipment urgently."

"Helicopter Three-Four. Message received. Wilco." Within minutes the great yellow bird was above, beating cold air on us. I kept up my work with the airway. Ian Nicholson was almost with us now.

The winchman spidered down from the chopper, spinning gently. Seconds later he was standing alongside us, unclipping his first-aid bag from the harness. He assessed the situation immediately, and quickly disconnected his harness from the wire rope swivel. Then, handing me the oxygen mask, he turned on the cylinder. Ian was with us now and we eased the lad into a better position whilst the oxygen filled his lungs. But, alas, there was no response — we were just too late. I felt sad, for had I been that bit quicker I might have saved him.

The helicopter had now flown off to avoid exposing us to noise and downwash and I suggested to the winchman that if they could take the body down to base we would try to find out if there were others in the boys' party that were injured. It wasn't possible for the helicopter to get close enough into Crowberry Gully above to find out. The walls rose several hundred feet on the right to the crest of North Buttress and on the other side, Crowberry Ridge, they were overhanging in places.

Ian had already set off up the gully and when Ronnie and Peter arrived, I continued up with them after we had put the body on the stretcher. We left the task of getting it hauled aboard the chopper to the winchman.

As we climbed up the frozen steps of Crowberry Gully, we could hear faint shouts from above, but owing to the bends in the gully it was impossible to see anything. We traversed out on to the North Buttress face to see if it was practicable to make contact from there, but we had no luck. Ronnie and I decided to follow Ian up the gully, whilst Peter elected to go down. He doesn't go a bundle on technical climbing, though he is very capable on steep terrain.

I contacted base, informing them of the situation, and suggested that they hold the rest of the team and the helicopter down below until we established whether or not anyone else was injured.

Ronnie and I were about half-way up the gully when we caught sight of Ian's tall figure above us. He shouted down:

"Hey, Hamish, there are four others up here. They're okay, but it will take a bit of time to get them down. I've another couple of climbers up here giving me a hand."

"Fine, Ian, I think we'd better see if we can find a place where the chopper can pick them up. It has a three-hundred-foot winch cable and may just get the end down to us."

"Aye, okay, but if we can't do that we could lower them all the way."

"The pubs would be closed by then, Ian, I'll try it this way."

As a couple of ravens croaked high above us, we found a place where a tongue of snow protruded out on the rocks on the face of North Buttress. It still looked very dubious for the chopper because the gully walls are very steep at this point, but at least it was possibly far enough away from the overhanging walls on the Crowberry Ridge side. I cut large 'bucket' steps across this steep slope, then excavated two ledges at the point where I thought there was a faint chance of getting the helicopter cable down to us. One ledge was for the first person to be lifted and the other for the next one in the queue.

Meanwhile, Ian was making good progress and they were now only a couple of hundred feet above. I contacted the helicopter pilot who had shut down at base and told him of the position.

"I don't think there's a great deal of elbow room for you," I told him, "but if you're willing to try, it could save a lot of work."

"Let us know when you're ready, Hamish, and we'll give it a go. The flying conditions are pretty good for Glencoe." Indeed they were. Usually, we have call-outs in foul weather with the helicopters thrown around like acrobatic kites.

I fixed a rope across the traverse from the bed of the gully and by the time Ronnie had secured his end, Ian, with his usual no-flap efficiency, had the first of the group down to us. I was amazed to see that they were all in their teens. We learned later that the youngest was fourteen.

"The pilot's going to give it a go, Ian. It'll save us some hassle."

"Great," Ian replied. "I could do with that pint."

The boys all looked shocked, as well they might. For, as we waited for the helicopter, I learned their story.

They were a group of Duke of Edinburgh Award youths, out for the day on their own, not on an official expedition. Arriving at the bottom of Crowberry Ridge, they had intended to climb up Easy Gully, an appropriately named climb requiring minimal technical skill. Coming to a branch in the path, they saw the well-executed stairway leading up the steep snow and mistakenly assumed this to be Easy Gully, when, in fact, it was the beginning of Crowberry Gully which, as I mentioned, had been trampled by Army lads only days earlier. With little hesitation the boys followed suit.

As they didn't have ropes, ice axes or crampons, they were soon in difficulties, but as the pitches get progressively more difficult in Crowberry Gully, they arrived at a point where they couldn't retreat. By this time they were almost at the Junction pitch, the point where the gully bifurcates. Here is the crux of the route: a traverse on steep ice, then a delicate climb often on thin ice directly upwards into the final lap of the gully, with the summit several hundred feet beyond. This pitch can often be technically awkward — for climbers properly equipped that is. For those young lads with not an item of climbing equipment between them, it was virtually impossible. They could not retreat, for below they had managed to struggle up several pitches which they could certainly never descend without falling. But to carry on would be equally dangerous.

They chose to continue upwards. The leader of the group had a bottle with him, which he broke and used to cut steps, using a splintered fragment as an ice axe. The others were close by, in a cave formed by the junction of the gully branches. To their horror, they saw their leader fall, careering down the steep snow. Then, after about 200 feet, he shot out of sight and went round the corner where the gully curves in towards the base of Crowberry Ridge. He had fallen about 800 feet. There was nothing for them to do now but to yell for help. Fortunately, the two climbers descending Curved Ridge, the other side of Crowberry Ridge, heard them and one ran down for help whilst the other stayed with the boy who had fallen.

Ian had found, when he got up to the four boys, that two further climbers were with them. One he knew as a fellow member of the Creag Dubh Mountaineering Club and who rejoiced in the nickname of 'Mini-Bugs'. The other lad was a stranger to him. They had been climbing the Gully behind the boys. Using their rope, Ian and the two climbers started to lower the boys down one at a time. The brother of the dead boy was in the party. Ronnie assembled the boys at his position at the gully-end of the traverse rope and I went back across to the ledges I had cut. I called the helicopter. Far below I could see our base truck and the yellow blob of the Wessex. In a space of a few minutes, the whine of the turbine could be faintly heard and then the throb as the rotors gained momentum. Ian and Ronnie had the first belay clipped on to the rope with a sling and a karabiner and Ian helped him over.

The shadow of the helicopter could now be seen above us on the rocks of North Buttress and presently the noise, ricocheting off the walls, became deafening as the chopper descended into the gully. There was very little room and I thought at first that the whole operation was a write-off, but slowly the pilot inched in. Now it was directly above our heads, I guessed about 400 feet. Still he eased it down and then seemed to stop. I could see the legs of the winchman dangling over the lip of the door. I knew exactly what he was doing. He was seated there whilst the navigator stood by the winch control. The winchman had already clipped on to the end of the cable and would have the wide strop attached to the swivel, ready to go over the head and under the arms of the person to be lifted. The winchman was now hanging from the winch gantry and was slowly descending. He was spinning more than usual. I didn't think there was enough cable for him to make it to the ledge, but we held our breath — there was, the 300-foot wire was virtually out.

The first lad, who I think was also the youngest, had a look of awe on his face. He must have thought this was real spaceman stuff. But he didn't have much time for contemplation. As soon as we had fielded the winchman on to the ledge, the strop was over the lad's head, brought up snugly under his arms. Ian whipped off his securing sling from the fixed rope and they were off, swinging and gyrating upwards towards that deafening noise. The operation was repeated for the remainder of the party of boys, then the cable came down on its own for the three of us. Mini-bugs and his mate prudently decided to climb down, concluding that their crampons and trusty ice axes were more reliable than that great yellow blob in the sky with its strand of gossamer.

Ian slipped the strop on. I don't know if it was the fact that there was only one person being winched up, or if it was turbulence, but at about 150 feet up in the air, he began to spin violently. He was quite unable to control it. Sometimes you can lean back with legs apart, acting like an air brake to stop spinning, but in this case his efforts were to no avail. Eventually, he was hauled aboard, feeling worse than he does after a session with my silver-birch wine. Ronnie and I followed and though we spun too, we didn't do as badly as Ian. Soon the glen rose up to meet us as the helicopter spiralled down to the car park. The ambulance was drawn up beside our rescue truck and a small group of sightseers had gathered. In the heather, unperturbed by the helicopter, a new born lamb greeted the spring, its plaintive call reminding me of the continuance of life.

FREAK OUT

AND

CHICKEN IN A BASKET

My ASSOCIATION WITH Alan Chivers stems back to the BBC Live Outside Broadcast of the ascent of the Old Man of Hoy, but his connection with television goes back 30 years. Alan, one-time ace Battle of Britain pilot and later spy plane pilot, is the pundit of live climbing TV shows. Though not a climber himself, he has an uncanny ability to pick the right camera angles to emphasize the fingergrip tension and edge-of-seat drama of climbing.

By 1980, he had yet to achieve his greatest 'OB', as they are called, the ascent of the North Face of the Eiger. I had suggested this to him in 1973, from a tent on the slopes of Mount Everest, and Dougal Haston, with whom I was then climbing, pooled his vast knowledge of the Eiger and his enthusiasm into the project. Chiv always felt that the Eiger broadcast would be done, but bureaucracy and procrastination have great appetites for time, and his retirement made its inevitable intrusion. As a going-away project prior to his retirement, he took on another climb, the ascent of a rock-face on Aonach Dubh in Glencoe called 'Freak Out'. I had interested him in this prior to returning to the Upper Amazon in late 1979.

Chiv had selected a successor to his seat at the BBC — Mike Begg. I have mentioned Mike earlier, he was going to make a film of our second Llanganati expedition. I had met him some time before in connection with film work, but it wasn't until much later that I got to know him well. Mike is tall, laconic and possesses a devastating wit. As a director, he is outstanding, always completely absorbed with the current production, be it a film or Outside Broadcast. His successes are diverse, and range from the Live Outside Broadcast of the Royal Wedding to the re-enactment of a dramatic mountain rescue on An Teallach in Wester Ross.

Mike can be a hard taskmaster if you happen to be an actor or a crew member, yet with climbers he seems very much at ease. He probably holds the record for the number of BBC secretarial desertions in a given year, but underneath this tough exterior lies a warmhearted iconoclast.

Things seem to happen to Mike . . .

One night he came home to his London flat to discover a burglar increasing his share of worldly goods at Mike's expense. Grabbing a poker, Mike made a Citizen's Arrest, and 'buzzed the fuzz' with his free hand. Meanwhile the burglar and the stalwart citizen stared at each other poker-faced, neither moving. The police didn't stir either, and after about five minutes the burglar sportingly said: "I'm afraid I'm going to have to make a run for it, you've had enough time." The burglar then bolted for the door, hotly pursued by Mike with his poker. The race continued into the street, with the pursued, a lean, whippet-like man, making headway. He shot round a corner at the precise moment when a squad car screeched up behind Mike and several burly figures in blue leapt out. Mike was thrown against a wall and quickly disarmed.

"I'm not the burglar, you idiots!" Mike screamed. "He's got away round that corner."

On the Glencoe 'Freak Out' Outside Broadcast, Chiv was to occupy a back seat, letting Mike take over the exacting task of directing the programme, where usually everything goes wrong!

Briefly, the set-up was that two parties were to attempt separate climbs on this overhanging rock face. Joe Brown and Jackie Anthoine, Mo Anthoine's wife, were to climb 'Freak Out'. Nearby, two of Scotland's leading climbers, Dave Cuthbertson and Murray Hamilton, were to do 'Space Walk', a 'new generation' climb, which would have been impossible only a few years ago. Not that 'Freak Out' was any afternoon stroll. When it was first climbed a number of years ago by a colleague of mine, Dave Bathgate, it was done in part as an artificial route. Now, the use of pitons has emerged as a sin akin to incest or high treason, and most of the pegs had been eliminated so that it is both strenuous and technically very hard. As a matter of fact, when Joe first tried it, he was put off by its reputation and gave up. Later he returned and climbed it prior to the programme.

An equally exciting aspect of a live-climbing Outside Broadcast is the technical side of things. Cables have to be run from each camera on the face to the scanner, situated at the nearest point of vehicular access. The scanner is an obese vehicle, the size of a large furniture van, which, with its attendant trucks, presents a formidable array. On this particular OB, satellite transmission was to be used for the first time by the BBC. This meant that a Ferranti unit with a dish aerial would beam the signal from the valley of Glencoe to the Eurovision Orbital Satellite. It also had a back-up system of radio links all the way to Inverness, via the Great Glen.

Everything had to be flown up to the bottom of 'Freak Out' face. This is where helicopter pilot John Poland came in — John is a tall, blond, ex-Fleet Air Arm pilot with whom I have worked for years. I spend a great deal of time flying with crack mountain rescue pilots both here and abroad, but John stands supreme. He can read every sign that an angry mountain offers, but as well as this he seems to have an extra sense, that edge which makes someone great in their particular field. He never gets flustered. As some vicious down-draught hurtles his Jet Ranger towards the ground, the only sign that I can detect when he has to concentrate is a little ditty which he hums and which comes over the headphones like the music they used to play to allay passenger fears in planes as they came in to land.

Some of the flying on this particular project was interesting, as John had to lay out five 4,000-foot lengths of cable with the helicopter, right up the steep cliffs beneath the climb.

It was the end of June, and Scotland had been sweltering under a heatwave. But of course the moment work on the programme started, the weather changed and it rained as if it was trying for the *Guinness Book of Records*.

Round the bottom of the climb, which was really a large ledge, we had various tents erected on scaffold platforms, and as there was so much valuable equipment up there, I had to appoint a resident guard. Peter Weir, a member of our local rescue team, had the first night, and Mike Begg made sure that his vigil would be tolerable with a 40-ounce bottle of Whyte and Mackay whisky in his survival hamper. As the guard tent was on the lip of a 300-foot drop, Peter, who doesn't have a good head for heights, tied himself down for the night, then contemplated his 'bottle'. Other guard nights were not without incident. In the interests of economy, Ronnie Rodger, also a team member, was given only a standard bottle for his night vigil, but Mike, realizing that mountaineers, like certain soldiers, 'march' better on full stomachs, arranged for John Poland to fly to the 'helipad' — a small tuft of grass alongside the tent — at 8 a.m. Ronnie was surprised to hear the helicopter so early, struggled out of his sleeping-bag, pulled on his pants, and rushed outside. John was hovering with one skid on the ledge. When Ronnie approached, he was amazed to see the rear cabin door swing open to reveal two fully uniformed, shapely waitresses stepping out to deliver a tray: porridge, bacon, sausage and eggs.

Despite the awful weather, the broadcast was a success, but owing to the severity of the climb, Jackie was still spreadeagled on the face at the end of transmission. Back at Broadcasting House, the switchboard was jammed by callers worried about her safety. However, Jackie was all right and got to the top shortly after the programme went off the air.

The previous night, Ian Nicholson had been on duty. He decided to make his bottle last all night, so he underlined each tipple with a pencil mark on the label. It was interesting in the morning to see how the marks had become increasingly erratic and wider spaced as the night progressed (the bottom line resembled a Himalayan skyline).

That evening was the last night for sentry duty on the mountain, and this time Ed Grindley was duty sentry. Ed was then a climber living in the Glen. Having heard the tales of nocturnal inbibing, Ed was disconcerted to find no bottle for his solitary sojourn. As I went down on the last flight, I promised that I would pick up some whisky from home and fly back up with it. With the helicopter engine running, I dashed into the house to find a roomful of people. Mike was there and Clive Gammon, a writer for *Sports Illustrated* magazine, New York. He had come to take part in a wager with me to balloon off the summit of Ben Nevis.

"Hi, Mike, Clive, I'll be back in ten minutes. I've just got to get a bottle of malt to our man on the hill."

"With the panic of transmission, I forgot all about that, Hamish," Mike said with remorse.

"No problem, Mike, as long as you replace my Talisker."

I didn't manage to join up with Clive, who had moved to Fort William, until two days later. The de-rig of the equipment on the mountain had to be completed and this meant spot-on co-ordination between John Poland and those on the ground — or face. Later the next day, I suggested to John over the radio that he could perhaps lift one of the thousand-foot lengths of cable directly from its top tether on the edge of the cliff at the base of the climb. It had been manhandled there with a lot of sweat and strong language.

"John," I continued, going into detail. "If you lower the chopper so that we can grab the wire rope on the helicopter lift hook, I'll couple it to the camera cable. If you then give your eggbeater full torque, I'll cut the tether to the belay, and presto — you can whiz back to base like a sea eagle with a long eel. If it's too much for the machine, you can always drop the whole lot, can't you?" I knew that there was a release from the controls to the lift hook.

"Seems all right to me, Hamish. Will we give it a go?"

"Right, John, next trip."

One of the other hazards for John, other than such suggestions from me, was Jaguar strike aircraft, which blasted through the Glen at 600 m.p.h. about 200 feet above ground level. John had informed Air Traffic Control that he would be lifting cables during this period, but I for one had fears of John's Jet Ranger plus BBC camera cable acting as an effective substitute for barrage balloon defence.

When the hook was attached to the end of the cable, I snatched my knife, which was ready-open on a rock nearby. John was about 60 feet above and there was still some slack on the wire hawser.

"Okay John, give it the gun."

"Right."

The helicopter gained height and shortly the boulder to which the camera cable was tied began to move. John was out from the face.

"Cutting now, John."

I just had to touch the nylon rope and it flew apart like a frayed whip end.

The heavy camera cable leapt out from the cliff and shot down and out some 30 feet as it came away from the face. The equal and opposite result of all this was to make the helicopter swing like an upside down pendulum.

"Jesus!" John said as he hurriedly gained control of the situation.

The other lads and I had a grandstand view from the top of the cliff and we were certainly worried for a few moments.

Later, when discussing this operation with John, I found out that he had assumed that I had used this technique in the past, which I hadn't, but he did in fact admit that it worked. Otherwise, as he says, "I would have come back and haunted you in bloody Glencoe."

With the pressure and the immediacy of the live-climbing O.B., I didn't have much time to speculate on the proposed balloon flight. It was only two days later, when a couple of heavily laden jeeps with trailers bumped along my drive, that the fact literally came home to me. . . . What the hell had I got Clive and myself into? Neither of us had ever put a self-respecting leg over the basket of a balloon before, yet here we were poised, so to speak, on the brink of the Ben Nevis precipices on a madcap flight!

The rules of the wager were simple enough: we would lift off from the summit of the Ben one after the other. The winner would be the first to land, after a minimum flight time of three hours, then pack up the balloon, run to the nearest pub within a 50-mile radius of the Ben Nevis summit and crack open a bottle of champagne. The Northern Constabulary, that strict guardian of Scottish Licensing Laws, had granted, after deliberation, permission for the aeronauts to partake of alcoholic refreshments at any hour of the day or night. Mike Begg's secretary, in turn, had contacted about 45 hotels and pubs, advising them of the competition and asking them to record exactly the moment when a glass of champagne was raised to the lips. The prize was a case of the stuff.

As the two vehicles ground to a halt outside my workshop, I reflected that this whole thing was the sort of wager concocted by undergraduates in an alcoholic haze.

Clive had meanwhile secreted himself in a hotel in Fort William and had bought 'high altitude' climbing boots from Woolworths, along with a selection of malt whiskies, for medicinal and, I suspected, anaesthetical purposes. Clive was even more worried than me! He is a Dylan Thomas-like man, and is indeed Welsh. Agile with a pen, he has the ability to prise out facts for a story without appearing to ask any questions, like a slick dentist extracting a tooth. He had been about to set foot in the Amazon with our merry band on our second expedition with Mike Begg and a posse of the BBC, but bureaucracy in Latin America decreed otherwise. I'm not sure who suggested going off the summit of Ben Nevis in balloons, but Mike Begg blames me.

Anyhow, Clive was contacted and, perhaps nudged by Bacchus, grabbed at the challenge, making a note in his diary at the time. Later, with clearer vision, he noticed this strange entry, 'Fat Bill' — marked for the month of June. Assuming that he was to meet Mike Begg at a restaurant of that name, Clive then spent some hours hunting through the Yellow Pages to locate it — without success. Eventually, from the hidden recesses of his mind, he realized that it should have read 'Fort William'. A short call to Mike in London confirmed his worst fears.

By the time he arrived in Glencoe, he had resigned himself to a trial of hot air and malt whisky. He was determined to depart both well-shod and well-primed with the back-up of a carton of tranquillizers.

Before I return to the arrival of our crafts and crews, I had better give some preliminary details. Obviously, you can't just hire a couple of balloons and attempt to float off the nearest mountain.

Air space is strictly controlled in Britain and it was essential to have *bona fide* pilots to assist us in this ploy, for both Clive and I were

Balloon taking off from summit of Ben Nevis

Capito, Joe Brown and the author with Lama helicopter

Outside broadcast: climbers on Freak Out

Five Days One Summer: doubles on the north ridge of Piz Palu

Ben Nevis: Mike Begg (in blue anorak) with the Royal Marine
Commandos digging out the tents at Camp Lazenby

Pontresina: Sean Connery filming in winter

novices. I knew very little about the art of ballooning; I had visions of rolling fields, lazy days, wicker baskets, wicker hampers and champagne. The stark reality was, we knew, cliffs, snow (even in June), incessant gales and cloud cover for 200 days a year, not to mention a mean annual rainfall of 162 inches. I didn't reveal these facts to Clive but, with his facility for gleaning information, he knew already. Actually, all he had to do was to mention the idea to any local to be told in no uncertain manner of the folly of such an enterprise.

My only other brush with ballooning had been when I had proposed using a balloon as a camera platform for filming an unclimbed pinnacle in remote country near the Green River in North America. Yvon Chouinard had spotted a huge rock spire whilst on a regular flight across Utah. He got a bearing on the virtually unknown area from the pilot, but future aerial investigation failed to locate this mysterious spike of rock. Meanwhile, back here in the UK, plans had forged ahead, but alas, the pinnacle never was found. Alan Chivers had also dabbled in balloons and had successfully used a military barrage balloon as a camera platform just after the last war.

We should have been further forewarned by Mike Begg's balloon preview. Mike is often tempted to partake in some of the 'stunts' which he films; perhaps it's the mark of a first-class director. I have recollections of him swinging dangerously from an overhang when there was some balls-up with ropes, or scrabbling up a greasy face after an all-night bender in some dank Welsh or Highland pub.

In this present instance, as an act of chivalry, he thought it expedient to try a flight himself before committing two of his friends to the heavens.

I had obtained the name and address of a balloonist from a colleague who was himself a member of the propane basket brigade. The pilot he recommended, whom I shall call Mr Know, was, he assured me, 'hot stuff'. Mike, being impressed by my implicit trust in my colleague's reference, arranged to have himself taken aloft above the tranquil English countryside.

Balloons, like fat ladies, don't appreciate a lot of wind, and the day of Mike's aerial baptism was just that. He told us of his experience when I returned from the mercy whisky mission on the night of the 'Freak Out' transmission. He had a captive audience.

"I once asked my uncle what it was like at the battle of Monte Cassino. He said nothing, he just cried a little. The memory of that bloody balloon flight brings on similar emotions. A strong wind was blowing as I arrived at the launching site, a field outside Milton Keynes which, if you've never seen it, is a rain-stained concrete dormitory-town looking not unlike East Berlin. I mention it because the pilot, in whose basket I was about to place my life, thought it one of the finest towns in Britain.

"The balloon canopy was strewn over the ground like a wrecked circus tent. Next to it stood the pilot and his helpers, casting anxious glances at the sky. Next to them were my flight companions, a small child who had won the trip as a competition prize and couldn't wait to get airborne, and a local bigwig who stood silently gritting his teeth and thinking of the free publicity.

"Our flight prospects looked bad. Balloons can only be launched in winds of nine knots or less and on that day even the pigeons were sitting with their backs to the gale. 'It is,' said our pilot, 'too risky to take the small child and the bigwig, but,' he added as I turned towards the car, 'we'll go anyway.'

"The balloon sprang into the air and everybody waved, that 'God preserve them' sort of wave normally seen when a troopship leaves port. We were one hundred feet off the ground and separated from it by a laundry basket suspended from an inflammable nylon canopy into which were leaping huge sheets of blue flame. Insanity is endemic to all balloonists; the basket, like a rabid bat, acting merely as the carrier.

"The pilot asked if I was nervous. I lied and, turning the map right way up, helped him locate our position. We were travelling fast and low and heading towards the dreaming spires of Northampton. Below us, cattle lumbered away, startled by the roar of the burner. Farmhands looked up and the pilot yodelled at them until they too lumbered away.

"The wind was getting stronger and the whole structure had adopted an unnerving shuddering motion. The pilot decided to land, which is when most balloon problems begin. We lost height and began to skim the hedgerows in search of a suitable landing site.

"'Every landing,' I was informed, 'is a controlled crash and in this wind the basket could be dragged for hundreds of yards.' In Bedfordshire, fields of that length are hard to find. The first suitable one, which had acres of soft green turf, also contained sheep. 'Can't disturb the sheep,' said this airborne St Francis, and he hit the burners and sailed over the field, leaving the gentle ruminants in peace and us, so far as I could tell, in the shit. We were coming in to land. There was a double barbed-wire fence at the rear end of the field and at the far end a bank of trees. Running the length of the trees and partly camouflaged by them was a high-voltage power cable. The pilot saw it at the same time as I shouted. He tugged on a rip cord, pulling a panel out of the top of the canopy. The escaping air sent the balloon crashing towards the ground where the first barbed-wire fence was coming fast to meet us. There was no room to duck down and our upper bodies were exposed. We still maintained tremendous forward momentum, enough to force us into the fence like two hard-boiled eggs going through a harp. I clung on and turned my face away. The basket smashed through the first fence and tipped over, tumbling down a ditch, up the far bank and rammed a second fence. This might have been made of rotten string, but in the event it probably saved our lives. We were still in the basket, and the partially deflated balloon was dragging us towards the power cable, but the momentum gathered by a balloon that displaces three tons had been slowed by impact with the fence which had acted like the arrester wire on an aircraft carrier. We were thirty yards away from the cable when the basket stopped. A farmer was running towards us. My hands were shaking as I crawled out but the farmer wasn't there to offer sympathy.

"'I thought you were going to hit that cable,' he snarled, and I nodded.

"'Good job you didn't,' he said, 'the power would have gone off and all my day old chicks would have died.'"

When we heard this, both Clive and I had butterflies which felt like hoody crows.

We exchanged glances, but it was too late to back out now.

"You know," Mike continued, sampling his favourite Grouse whisky, "there was some Earl last year who was fined for firing his shotgun at a balloon that was disturbing his shooting party. He missed both the balloon and its occupants, but it proved that even though the aristocracy's marksmanship is in sad decline, its sense of values remains intact."

The 'crash' had not endeared Mr Know to Mike, though he found Paul, the other pilot, a professional, quiet and unassuming.

I watched Clive Gammon swing into the drive in the wake of the jeeps for this gathering of the clans. When the bulgey jeeps pulled up, I was surprised to see two girls tumble out, apparently the pilot's girlfriends, Priscilla and Joan.

The weather had cleared and it was balmy spring again. I forgot all about the water-cannon rain of the past week and almost looked forward to this new project.

It didn't take Clive and me long to discover that Mr Know was an avid lecturer. Even the simplest question from either of us resulted in protracted dialogue, but we did discover that the balloons were to be powered by double propane burners rather than the usual single unit, and that I was to fly with Mr Know whilst Clive was to share Paul's basket. The prospect of landing in one of the mountains which surround Ben Nevis didn't bother me much, but I did have a fear of landing in the sea, and I asked pertinent questions in respect of balloon buoyancy and the availability of life jackets. I knew one didn't last long in the cold waters of the North Atlantic, having little faith in the so-called 'warm' water of the Gulf Stream. My cowardly questions betrayed a hydrophobia and gave Clive a new lease of life. He didn't mind water (except in whisky), and his Achilles heel, 'cliff-phobia', had already been revealed by the purchase of Woolworth's £15.00 'high altitude' boots.

"You've nothing to worry about," Mr Know assured me. "You can always hang on to one of the empty gas cylinders."

"And," Mike Begg cut in, "there are going to be two helicopters flying about anyhow." One was for the BBC cameraman and Mike, and the other for the *Sunday Mail* newspaper, who wanted to do a feature on this mad competition.

"What if we land on a cliff face?" Clive cut in, taking a nervous drag on his cigarette. He had an uneasy look in his eye and glanced surreptitiously at the ominous west wall of Aonach Dubh, just up the Glen.

"That could be a problem," Paul replied, "but usually we manage to get down on flat ground if there is any."

I thought I heard Clive mutter that there wasn't, but I was not sure. Mike promised to do something about the contestants' phobias. We were both to be equipped with crash helmets, for Clive's projected cliff hang-up, and life jackets for my North Atlantic dip. Mike managed to borrow these from the RAF Station at Lossiemouth.

I was concerned that the makers of the life jacket were 'Frankenstein and Sons, Manchester', and Clive, in turn, worried that his helmet "was inscribed with the names of the previous owners, each one carefully crossed out, so that it resembled a small war memorial".

Play moved to 'Fat Bill'. The weather couldn't have been more considerate — clear skies, and a gentle zephyr wafting sulphur-dioxide emissions from the nearby pulp mill over the town. The 'When birds do sing, hey ding-a-ding' weather didn't suit Mr Know, who was the technical head of the operation. He studied his anemometer as if it were a crystal ball which featured the Armageddon as a coming attraction, and let off small helium wind-indicator balloons like offerings to the Aeolian gods. These rose gracefully in the pleasant breeze.

Perhaps reflecting on my earlier fears of landing in the Atlantic, Mr Know now raised this as a distinct and dangerous possibility, but with a westerly wind blowing, I found the suggestion difficult to comprehend, as it would be the North Sea which would be the grateful recipient. Anyhow, the helicopters were ordered for the next day and we returned to our various refuges to gird our loins for the morrow.

A helicopter arrived the next morning expecting to shuttle the folded balloons in their baskets up to the summit of the Ben, but instead sat waiting whilst there followed more professional dialogue about wind. A recce to the summit was decreed essential to release more helium balloons.

Arriving on top, we surprised flocks of tourists, who looked on in amazement as Mr Know launched his 'bubbles' into Scottish air space. It was a scene reminiscent of that renowned advert for Pear's soap, or perhaps that well-known illustration of Casanova demonstrating one of the secrets of his popularity.

Upon returning to base, the verdict of the professionals was 'marginal'. Someone was heard to mutter something about 'chickens in a basket'! However, the scales were to fall on the dangerous side of 'marginal' with the later arrival of a north-east wind, which swept down the Great Glen. Back to the pub, where our training diet of orange juice and cider was discarded in favour of Highland malt.

Clive and I felt that we had squandered some of the best possible weather, and I was also aware of the fact that, once the north-easterly wind starts, it often forgets to stop.

"We would be lucky," I told Clive and Mike, "to get a long enough break to launch our 'chicken' baskets."

To add to all this, we could only keep John Poland and his helicopter for three more days, as he had another assignment lined up — to take Rod Stewart on a fishing trip.

'The Mets', as our professionals described the forecast, were avidly studied. When Clive heard this, he thought they were discussing the form of a well-known American baseball team.

It seemed that Friday the 13th, two days hence, was our only hope. Everything was prepared to be flown to the summit the following evening, if the wind abated.

Ian Nicholson was summoned from his retreat in Glencoe to assist and guard the equipment on the top. The press arrived from Glasgow, weighed down with duffel coats and cameras and the cynical expressions of men about to witness attempted suicides. I felt that nothing short of the contestants' blood soaking into the peaty ground of Lochaber would satisfy them. Their helicopter was being driven by John Poland's partner, David Clem.

It doesn't really get dark in late June in Scotland, and the equipment was flown to the summit through early evening without a hitch. Ian went up with it, taking his sleeping-bag. I was sorry that I couldn't join him, for it was a stunning night. He later told me that there was enough light to take photographs at 2 a.m.

The next morning at three o'clock, Clive and I arrived at the landing site in the conifer forest at the base of the mountain. There was no sign of the balloon pilots. I had said to Clive: "I don't think they're coming," when we heard the throb of the jeep engines shortly obliterated by the high-pitched whine of two Jet Rangers spiralling into the clearing.

Within an hour, everyone was on the chilly summit, including press and crew. The weather seemed perfect, yet operations were held up for a further two hours for no apparent reason. Both Clive and I were wanting it over with, feeling the impatience of men too long on Death Row.

Eventually the balloon was laid out on the snow and hot air was blown into the envelope. First, it was my balloon, for I had won, or lost, the toss, whichever way one interpreted it.

It was the sort of day which the Scottish Tourist Board should preserve to delude would-be visitors. Peak upon peak strode to the horizon and I saw the cold metallic blue of the Atlantic to the south of the jagged Cuillin of Skye. Indeed, there seemed to be sea and lochs everywhere to the west with only the dark spits of land fingering this watery mirror.

Clive probably read my thoughts and rubbed salt into an already tender spot.

"This easterly wind could take us down into the Sound of Mull, Hamish, or possibly north of Iona. I suppose the next land after that is Labrador."

"We have one or two mountains to get over first. I don't think either of the pilots go a bundle on a salt-water touch-down." Clive didn't reply but lit another cigarette.

Priscilla and Joan were throwing snowballs with much shrill laughter and inaccuracy. Mr Know, now that things were in a 'go state', was introspective. There could be no more excuses, he was on the brink — so were we! Pressure was building up, both in the envelope and with the presence of the media. Cameras were rolling, and the huge ungainly contraption came to life (the balloon, not the media). We were in the lee of the summit, and when the balloon reared its head into the breeze, it became agitated, shook, and attemped to free itself from its nylon tape leash.

We had been given the balloon safety drill a hundred times over, like two children having to learn the Catechism:

> "Where is God?"
> "God is everywhere.
> He watches over us with loving care . . . "

I certainly hoped so.

Gloves were on, Frankensteinian life jacket, war memorial helmet.

"Stay in the basket at all costs."

"Let the basket absorb the impact." We should have known when we first heard that one!

We were aboard. It was no longer the cissy fore-and-aft sport I had imagined. The two burners were roaring inches above my head, huge inside-out umbrellas of blue flame. Big Ian and other casual volunteers were attempting to take some of the strain off the tape. Mike didn't have to shout 'action'; we were off whether we liked it or not. The nylon tape, with a breaking strain of over 3,000 lb., snapped, and we were dragged across snow, then rocks, towards the abyss which contained Glen Nevis 4,000 feet below. We were airborne — I could hardly believe it. Now lonely as a cloud, we drifted westwards, relatively peaceful except for the blast of the gas burners.

Meanwhile, back on the snowy summit, Paul started to inflate the other balloon just after we broke loose, and this was now doing its ritual dance on the end of its lines. The cheers which had accompanied our 'launching' had almost died away when Priscilla, who was herself a balloonist, shrieked, "Oh, Jesus!"

The reason for this 'please do something about it' exclamation was due to the fact that up in our basket, things went suddenly wrong, or rather, continued to go wrong, for the balloon began to fall at an alarming rate. Clive described it as "like two gigantic hands which slapped the sides of the balloon so that it collapsed inwards". Paul stopped work on his balloon, and ashen-faced, muttered 'wind sheer'. This is the term for rogue gusts, many of which apparently live round exposed mountains. As we fell towards the ground, I was contemplating the feasibility of jumping upwards at the moment of impact, but the possibility of being cremated on the burners had even less appeal than doing an Easter-egg roll from the south-west shoulder of the mountain.

"Give me a hand, Hamish," my pilot shouted. He was attempting to close the parachute. It wasn't wind sheer which had caused our disorder, but 'a minor error', as he described it later.

"I should have broken out the parachute before take-off for more lift, but I did it after we were up, and lost a great deal of air." We certainly did!

We lost so much air that the flameproof material of the tube leading from above the burners to the envelope collapsed, and the burners weren't shovelling air into the balloon at all. Meanwhile, we had dropped 1,000 feet.

Eventually, the massive burners took effect and the huge envelope again billowed into a more reassuring shape. We stopped falling. Both our faces were coated with sweat — of the cold variety. Even in my ignorance I realized that it had been a very close call, and reflecting on the matter later, realized we were lucky to be still alive.

After all, no one had ever taken off from the summit of Ben Nevis in a balloon before, and you would think this an elementary precaution.

As we disappeared from view, those on the summit had concluded that it was probably the premature end of two aviators, and Mike Begg was in a quandary. For when he saw our balloon disappear rapidly from view, he decided to sprint to one of the waiting helicopters to attempt to film the impact — at least, he concluded, it would make good news footage. However, just as they were scrambling aboard John's chopper, the drama shifted to Clive's balloon. The inflation of the other balloon was still under way and this was now threatening to break loose. Paul got aboard and Clive had one leg over the edge of the basket when a gas leak ignited and burst into flames. Then it broke one of its tethers. Fortunately, this time there were two tapes being used instead of our one. Clive hurriedly retrieved a singed leg and reeled out of immediate danger. 'Better,' he thought, 'to be chicken without a basket than a roasted one.'

Paul, inside his wickerwork oven, reached outside, grabbed the fire extinguisher and managed to kill the fire. Had one of the large propane cylinders blown up, the summit party would have been sadly depleted. Quickly he released the air from the balloon, and in a minute the drama was over. Paul was singed but otherwise unhurt, probably saved by his flying suit and helmet.

The result, as far as Mike, his film unit and the press were concerned, was that they missed out on both episodes, except for some footage when our balloon disappeared from view like Halley's Comet.

Meanwhile we had relayed our deliverance to Mike over the walkie-talkie and he in turn told us of Britain's highest pyrotechnic display. Presently, John Poland's Jet Ranger swooped down and they filmed our stately envelope lazing above Glen Nevis. Mr Know was reluctant to venture any further to the west and water, and I grew somewhat impatient with this state of suspended oblivion. It was rather like dozing in the dole queue.

After 80 minutes, he let just enough air out this time to proceed to the ground more sedately, and after boulder-hopping a stone dyke, we dragged along the peaty hillside at the mouth of Glen Nevis, appropriately, I thought, stopping with a thud on the site of the Battle of Inverlochy, where the valiant Montrose vanquished the Covenanting army. Just ahead of us was the main high-voltage line for the region.

By this time, Clive, Paul and the rest of the summit party had been flown down and, after we had packed up our balloon, we walked to the nearest pub, the Nevis Bank Hotel. Here, over champagne, Clive, Paul and Mike joined us. My pilot raised his glass (after returning the first bottle of champagne which wasn't to his satisfaction), and he stated, "It was not an epic flight."

I nodded in agreement.

Then he continued, "Just a great flight."

"At least, Clive," I said with raised glass, "it's our last flight."

FRED CALLS IN THE

MAFIA

ONE FINE AUTUMN morning in 1980 I received a letter from an old acquaintance, Norman Dyrenfurth. Norman and I had worked together on Clint Eastwood's *The Eiger Sanction* and had since stumbled across each other in remote places. In his letter, he mentioned the possibility of another mountain film, this time directed by Fred Zinnemann and produced by the Ladd Company (of Alan Ladd Jr fame).

I had heard of Fred Zinnemann, of course, but not being a film buff, I could recall only one film of his vast repertoire, *High Noon*. It transpired that Norman had known Fred since his youth, but their paths had never crossed professionally until the germination of this latest film.

Peter Beale had already been appointed Executive Producer of the movie, which had a working title of *Maiden-Maiden*. When he spoke to me on the phone a short time later he suggested an early meeting in London with Mr Zinnemann. Even over the telephone, Peter conveyed an almost boyish excitement about the project, this being tempered by a knife-edged business acumen.

"It's a big project, Hamish," he told me, "and we want you to take charge of safety, as well as advising. Norman will be running the Second Unit."

"When is it due to start?" I asked, thinking that it could perhaps interfere with my hardy annual expedition jaunt, though just then I didn't have any plans pending.

"We would like you to be available on an advisory capacity as from now, but shooting doesn't start until next year, probably June 1981."

"That seems fine by me, Peter. Anyhow, we can discuss it when I come down."

"Right, Hamish," Peter returned briskly, "I'll look forward to seeing you."

When I met Fred Zinnemann at his Mount Street office, my first impression was of a man fragile and delicate. This immediately made me apprehensive at the thought of his going up high mountains. However, my doubts quickly evaporated once he spoke: his frail exterior concealed a will of iron. He reminded me of a withered pine branch which had aged over the years and now resembled spring steel. Longfellow's lines from *Excelsior* seemed to depict him:

> Beware the pine-tree's withered branch!
> Beware the awful avalanche!

I chose not to dwell on the second line, we didn't need bad omens!

Fred must have sensed my initial doubts, for one of his first remarks, after formalities had been dealt with, was: "The doctors have given me the all-clear to work up to fifteen thousand feet, Hamish." His eyes twinkled and he was clearly bubbling over with enthusiasm.

I discovered that he had spent his early years in Vienna and had indeed walked and climbed in the Austrian Alps. This was obviously an important period in his life, for mountains had never lost their fascination for him, even during his long exile in America. In his prolific career, he has directed 22 feature films, many of which are now household names world-wide, including *From Here to Eternity*, *The Nun's Story*, *Julia*, *A Man for All Seasons*, and not forgetting *High Noon*. I was surprised to hear later that he had also directed *Oklahoma*, which, I felt, after meeting the man, was out of keeping with his style. Over this creative half-century, he was awarded two Oscars as well as countless honours such as four Scrolls from the New York Critics' Circle.

"How," I asked him, as I sipped a cup of coffee, "did you get this idea of making a feature film set in the mountains? It's surely the most demanding environment possible for such a venture."

"I've always had a hankering to make such a movie, but I never did find a suitable plot. It wasn't until I read a short story by Kay Boyle called *Maiden-Maiden* that I felt I had found the basis for a feature film. Michael Austin, the scriptwriter — you'll meet him presently — wrote the screenplay based on Kay's story. You've read this, haven't you?" he asked, holding up the script.

"Yes, I think it's good. There are only a few small points, technical ones, which need tidying up."

"Well, we'll be relying on you and Norman to put us right on such things," he returned.

"The action is a bit before my time — I think Norman will be more help there."

In fact, I had read the script the previous night on the sleeper to London, and realized the scale of the enterprise. It would all require careful and detailed planning. A lot of the locations were to be high in the Swiss Alps, possibly in the Pontresina–St Moritz area. This is a beautiful part of Europe, with a diverse selection of glaciers and peaks, from the snow-draped Piz Palu to the tombstone granitic majesty of Piz Badile in the Bregaglia. It was, we were later to discover, the ideal location.

Not only were these high locations going to present headaches, but the scenario was set in the early 1930s, when mountaineering equipment was very basic and the climbing of steep ice hazardous, to say the least. This was due to lack of suitable belays. It wasn't until much later that ice screws were invented for tying on to, and ice axes emerged which stayed put when you wanted them to. It was a physical era, with step-cutting in full swing and the use of crampons just beginning to make an impression.

There was a knock on the door and a tall, lean man entered. Mr Zinnemann introduced me to Michael Austin. Michael had the easy-going manner of someone used to the outdoors and indeed he is an accomplished yachtsman.

"I'm glad that you're working on the film, Hamish," he said. "I should think we've given you a few problems."

"It looks like it," I replied with a laugh. Michael apologized for not being able to stay, but he had another appointment.

"I liked the script," I told him as he took his leave, and I stood up to go myself, for I had to return north that afternoon.

Before leaving, I said to Fred that I would start working on my side of the film and compile a list of the climbers whom I would need for safety. He was quite familiar with the big names in the climbing world and was delighted when I mentioned some of those who would be working with us.

Before I left London, I met Peter Beale at his London home. Walking in the front door, I was intimidated by the blood-red walls of the interior and a profusion of pillows covering the settee, rather like those in a sheikh's tent.

But Peter himself looked as I had imagined him over the phone, reasonably fit, with a boyish face. I learned later that he was associated with the production of such films as *Star Wars*, *Alien* and *The Omen*. He had grown up in the film world, starting his career when a boy at Pinewood.

"I've no plans to go off anywhere at the moment, Peter, so I'll start work on equipment — the special ropes and suchlike. We'll need simulated hemp ones for the climbers but they'll be proper climbing ropes. I'll also make you out a list of addresses for other specialized gear which we'll need."

"Have you got a safety crew in mind, Hamish?" he asked. Then he added, "We'll have to include some local Alpine guides to keep matters sweet with the Swiss."

"Sure, I'll get that in hand. Fred Zinnemann mentioned that he thought the Bernina Range was high on the list for location work."

"That's right, do you know the area?"

"Yes, I know it. I haven't climbed there, but it seems a good choice to me. The weather's better there than most other places in the Alps." I leant back on a couch amidst a cluster of cushions. "By the way, have you cast any of the actors?"

"Not yet, but we hope that Sean Connery can play Dr Meredith."

"And the guide?"

"That's going to be difficult, don't you agree?"

"I do."

The basis of the script is a love triangle involving a Scotsman, Dr Meredith, holidaying in the Swiss Alps, his attractive niece — posing as his bride — and a young Swiss guide whom they hire. Dr Meredith is a climber, as is the girl to a lesser degree. The incestuous relationship with the niece, Kate, is unknown to the doctor's wife. Although basically simple, I felt, the love story had great potential under the magical direction of Fred Zinnemann.

Recruiting for the safety brigade wasn't difficult. I had already tentatively asked various climbers. These were mainly the well-seasoned veterans, who had worked with me on many films: Joe Brown, Martin Boysen, Mo Anthoine, Paul Braithwaite, Paul Nunn, Rab Carrington and Ian Nicholson ('Big Ian', as he is known in climbing circles, who is also a member of the Glencoe Mountain Rescue Team). Later on this nucleus was to be supplemented by other well-known British climbers. Together they were dubbed my 'Scottish Mafia', a dubious title which became the accepted alias of the safety squad.

Over the next few months I met others concerned with the movie-making machine in London, and we thrashed out plans for operating at altitude; much of the camera equipment had to be specially adapted. Sean Connery had by this time agreed to play the part of Dr Meredith, and we discovered that Paul Nunn looked a convincing climbing double for him. Ian Nicholson was another possibility, so to begin with Ian would play in the long field in case Paul had to return to his lecturing commitments at Sheffield Polytechnic, where he taught economics.

Arthur Wooster was appointed chief cameraman of the Second Unit. Arthur is a small, bespectacled, innocent-looking man who has a lot of bottle. He had done much of the special Second Unit work for the James Bond series. The second cameraman in the Second Unit was Herbert Raditschnig, an Austrian and a leading Alpine climber of the last decade. Assisting them with camerawork would be Tony Riley, an expeditioner with the patience of Job, who would dangle for days in order to capture an idyllic moment on some great mountain wall. As there seemed to be a surfeit of cameramen, it was decided to have Leo Dickinson, with his mountain partner, Eric Jones, make a documentary of the whole operation. They were a talented, well-tried combination for such a task, used to working on their own. Leo has a wealth of adventure-type documentaries to his credit, such as the ascent of the Eiger and Matterhorn North Walls, and other equally dramatic works on the high-angled stages. Eric was no mean mountaineer, either. He had, a short time before, made a solo ascent of the Eiger North Wall.

Though the film had a relatively small cast, Mr Zinnemann still hadn't found an actress to play Kate. She had to be fit and, even though a double would be available, she would obviously still have to do a certain amount of mountaineering. Also, most important of all, he needed an actor to play Johann the guide. I had suggested Terry Kingly to Zinnemann. Terry, as well as being a climber of the upper echelon, is also an actor, but Terry was not what Zinnemann had in mind for the character. Nevertheless, he chose him to play the part of a guide in the rescue scene. In April, Zinnemann auditioned a young French actor called Lambert Wilson and was now seriously considering him for the part. Eventually, he decided that he could do it. When I heard of this from Peter Beale over the telephone, my first reaction was "Can he climb?"

"I understand he can, Hamish. At least his agent says that he has done a number of big Alpine routes."

"That's great," I enthused. "Shall we have him checked out over here? I can take him on a climb in Scotland, or if time's short, Joe Brown or Martin Boysen can fix one in North Wales."

It was arranged that Joe was to take him rock climbing over a weekend in Llanberis. A hotel was booked for him and Joe managed to borrow an extra pair of climbing boots. However, Lambert was tied up in some dance production in Paris and the visit and test never materialized. As an alternative it was scheduled that he should have tuition on climbing techniques of the thirties in Chamonix at weekends with Jean-Franc Charlet, a Chamonix guide. Jean, who comes from an old family of famous French Alpine guides, is a friend of Rab Carrington (one of our climbers); he later worked with me on Ben Nevis in a live outside broadcast where he skied down the intimidating north-east face of that mountain.

A few weeks later I suggested to Peter that Lambert's progress should be checked, and though I couldn't go to France myself I suggested that Martin Boysen should. Imagine my surprise when I spoke to Martin the following week when he returned. I had phoned him at his home in Manchester.

"He's doing as well as one can expect, Hamish. He can abseil and use crampons reasonably well; he's also done a little on rock."

"But I understood that he was an experienced Alpine climber!" I said with astonishment.

"Oh no, he's a raw beginner, but he's not doing too badly."

Peter Beale was also at a loss when I relayed this intelligence to him, but there was little we could do about it now; after all, Fred Zinnemann thought that from the acting point of view Lambert would be fine.

In London I subsequently met Andrea Florineth, the chief guide for the Pontresina area. As well as speaking perfect English, he seemed an easy-going person, which was just as well, as he was to be our key man out there. I took an immediate liking to him. He had the relaxed attitude of one who wasn't easily ruffled, and with a surplus of ability. He also had the advantage of having worked on films before. In fact, he had doubled for James Bond in one of the early sagas, so he was conditioned to the interminable hanging about which goes with making feature films and which climbers find extremely boring. He knew the Bernina-Bregaglia area like his own bath-tub and this knowledge was to prove invaluable. Half the battle with mountain filming is finding easy yet realistic locations.

I also met Freddie Francis on one of those visits and was sorry that in the end he did not take on the job as Director of Photography, but Giuseppe Retuno (nicknamed 'Pepino'), who was eventually persuaded to join the happy gang, was a wonderful choice. A charming man, with close-cropped black hair and just a peppering of grey, Pepino often looks as if he has just spilt a glass of vintage wine. This analogy completely belies his lighthearted zest for life. But at work he is totally dedicated and absorbed, sometimes with his head in the clouds and sometimes his feet and legs in a crevasse — how we didn't lose him down one of the bigger hidden crevasses I don't know. His enthusiasm was contagious, even in the worst weather. With Pepino came a complete ready-for-work Italian camera crew; several of them couldn't speak any English, but they proved just as endearing as their Maestro.

New personnel were being recruited every day and it was difficult to keep track of them all. Tony Waye was appointed Assistant Director to Mr Zinnemann, and Art Director was Willie Holt, a talented Frenchman who, with a small legion of French tradesmen, could fabricate an Alpine hut and even a hotel in a remote valley in incredible detail and in an astonishingly short space of time.

The rôle of Kate still hadn't been cast. Mr Zinnemann had interviewed dozens of actresses, but none came up to his expectations. He was, I discovered, very fastidious when it came to casting and, for that matter, in all aspects of film-making. Each actor had to be physically and mentally suited for his (Zinnemann's) interpretation of the rôle. This is what probably makes him one of the great directors — he never accepts second best.

He had now firmed up on the location for shooting and it was, as expected, the Pontresina area. It was proposed that some of us should go out during May for a week prior to the main crew arriving in June to check locations. Fred Zinnemann had visited the Pontresina area on other business in 1979; it had occurred to him then that the locality would be suitable for making a climbing movie. Norman Dyrenfurth, who did some early surveys with Fred, also favoured this part of the Alps. But Fred's recollections took him back to before this, when the glaciers were reaching further down the valleys. He had climbed there as a young man in 1929.

Zinnemann was late arriving in Pontresina owing to leap-frogging between New York, Paris and London. He had narrowed the field of the leading lady down to two actresses, one in New York, an English girl, and one in London, an American. His Paris visits were to confer with Lambert Wilson. At the tenth hour, his choice for Kate was Betsy Brantley, the American who, at that time, was playing a part in a London musical. We learned from Zinnemann that she was from the Blue Ridge Mountains of North Carolina and, as well as being brought up with a feeling for the hills, was one of the modern super-fit young ladies. Fred thought the part suited her. She was an accomplished, though relatively inexperienced, actress. As with Lambert, this was to be her first major part in a feature film.

There were some priority questions to settle on this pre-shoot: glacier locations had to be reasonably close to the madding crowd, for we seemed to resemble Fred Karno's army. Even at this stage I was having kittens about the numbers scheduled to work on the glacier. I just couldn't see why 120 crew were required for relatively simple sequences, involving only three actors, with the exception of one or two scenes.

My thoughts took me back to the Clint Eastwood film *The Eiger Sanction*. I had then made a ruling that only climbers should go on to the Eiger North Face — other than Clint. This precaution was well-founded, for before I took over, a friend of mine, David Knowles, had been killed in a rockfall. Indeed, on the Eiger the 'small is beautiful' policy worked well. On *Maiden-Maiden* it was difficult to discover why such a gargantuan crew had been put on the payroll; it was probably due to union machinations. Both Peter Beale and Fred Zinnemann were willing to work with a more compact group, but despite my suggestions, the paradox of the 'safety in numbers' policy remained, which caused us some nail-biting over the next few months.

None of us underestimated the problems of working with the local Swiss guides; they are not the most understanding people on this planet. I could, to a degree, sympathize with their position. After all, they had grown up in those mountains and had a right to feel pissed-off with so many blue-jeaned British moving in on their patch. Still fresh in my mind was the trauma of the live TV Outside Broadcast from the Matterhorn some years before when we did a televised centenary climb of the mountain. Then the project was almost sabotaged by the Zermatt guides. Fortunately, the Engadine guides proved to be more amenable. Andrea Florineth, the boss guide, was a godsend. Not only did he understand the need of British climbers to be working on the film (after all, no one could doubt their qualifications), but he proved later to be an astute diplomat when dealing with tricky situations.

I met the two pilots who would be working with us, Ueli Baersuss and 'Capito' Wismann. Both were experienced rescue pilots and I knew some of their colleagues in Air Zermatt with whom I had flown on the various Eiger North Face recces.

Ueli is a thick-set, archetypal stolid Swiss, who has been flying rescue missions for donkeys', or perhaps one should say eagles', years. He is unflappable and utterly reliable, yet he has the hair trigger reaction of a mongoose. Being a climber, he understands mountain situations and later, when I got to know him better, he used to park his helicopter after we disembarked on some lonely Alpine ridge or buttress and head after Andrea and myself on our many survey missions.

Capito was the antithesis of Ueli; he was younger and hyper-active. When we were working on the film, he used to do the famous 'Capito dive', hedge-hopping the Lama or Alouette over a ridge to fall like a stone down the other side. Not only was your heart in your mouth, but everything else was as well.

On the day after we arrived in Pontresina, Andrea suggested that we should both have a look at the glacier on the east side of Piz Palu, which he thought was a possibility for the main glacier scenes. Mr Zinnemann had seen it previously from the air and thought it looked promising. With us came Alfred Schmidhauser, the oldest active guide in the valley, a charred, cherubic-faced man with short, springy scrubber-brush hair and a 24-hour smile. At seventy-five he still drove a sportscar with panache and had the energy of a ginseng addict. He was to be with us for the duration of the film and was later enrolled as an actor, Johann, the hut guardian, which he did very convincingly. He had in truth been in charge of an Alpine hut at one time.

Anyhow, when Andrea and I got out of the helicopter on this remote glacier to 'eyeball a crevasse', as my American friends put it, Alfred, for some reason best known to himself, ordered the pilot to take off whilst Andrea and I were peering into the blue-green depths, leaving us stranded in this dangerous place with not so much as an ice axe between us. Our rucksacks were still in the locker of the machine. However, the pilot, discovering that we had not intended to be low-temperature castaways, returned half an hour later, much to our relief!

We didn't have much luck on that first survey looking for glacier/crevasse locations. One possibility, the Morteratsch Glacier, had been inspected by Fred Zinnemann during the winter when I was in South America. This massive glacier has a snout which spews down from the Piz Palu like a great white-bearded grub. When Fred studied it months before, it was pristine white — now, in the youthful light of spring, it had turned several shades of grey, with promise of a dirty black hue by midsummer, when we planned to work on it. In addition to this obvious problem, despite diligent scouting we failed to find a suitable crevasse. This was for a scene where a body, interred and deep-frozen for 30 years, was to be found by the guide. Andrea took us to a crevasse which he himself had fallen into when solo skiing a couple of years before. He was very lucky to have escaped with his life, for he lay in its icy depths for eight hours before being discovered, but it was not what Zinnemann wanted.

Despite all of these apparent failures to pin-point locations, I think we realized that this region had everything to offer; it was just a matter of finding the right spot for the right sequence.

About this time I met Leonard Gmur, the Pontresina Location Manager. A straight-backed Swiss, Leonard has a Scrooge-like record for saving film companies countless dollars by astute management. He reminded me of a Prussian count, and he did indeed wear high leather riding boots and a gaberdine raincoat draped over his shoulders with the arms dangling at his sides like an amputee from some cavalry campaign. Leonard was a key figure in this vast film-making machine.

I had suggested to Peter Beale that training for actors and crew was essential to safety. Peter went along with this, despite the enormous cost.

There was no question of using stuntmen for the climbing scenes, as Alpine climbing expertise of a high order was required, and this talent — in the form of doubles — came from the ranks of the 'Mafia'. Paul Nunn and Martin Boysen provided such realistic facsimiles for Sean Connery and Lambert Wilson that it was difficult to tell who was the double and who the actor from 20 feet away when they were in costume.

As one of the scenes was set in a Clydeside shipyard, Betsy was sent, that May, to visit Ferguson's of Port Glasgow, the shipyard which was to be used for this scene. In this way she could get the feel of the yard where she was supposed to work in the drawing office.

Though the mustering of the film crew wasn't until June, I set off at the end of May to drive to Pontresina. Over the next few weeks the crew would multiply and I had to ensure that the training programme would be operational on their arrival.

On the way out I called at Ehrwald in the Austrian Tyrol to visit an old friend of mine, Hans Spielmann, the owner of the Spielmann Hotel. I had been fortunate in spending eighteen months of my National Service stationed in Ehrwald, which was then a holiday camp for the British Army of the Rhine, and as it was in the then French Zone of Austria, discipline was at a minimum. I spent a great deal of off-time climbing those fairy-tale mountains with Hans as my constant companion, and we built a strong and lasting friendship. His father, Christian, who had been a well-known guide, was a gnarled old man with a shock of white hair and deep-set brooding eyes which always seemed to be focused on the high tops as if he was constantly reliving his youth.

The last time I had been to the hotel was in 1970, when I was returning from a Caucasian expedition. Paul Nunn had been with me then and the ensuing party had left our departure from this tranquil place vague in my memory. I know that Paul forgot his jacket and Arnold Larcher, a well-known Austrian climber who, as well as being a guide, was also the local baker, had suffered the wrath of the populace. On the morning of our departure, the good people of Ehrwald had been deprived of their morning rolls, as Arnold had been involved in our nocturnal celebrations.

Arnold showed up on this present fleeting visit, along with many of the local rescue team. We drank to old times whilst Arnold played on his zither, a recently acquired passion. He had just returned from an expedition to the Cordillera Blanca of Peru and was preparing to join a German expedition to the Himalayas to attempt a new route on Makalu.

If anything can make the Swiss smile, it's the fact that a film unit is moving into town for three months. Though the many hotels in Pontresina and St Moritz are booked up year after year, I'm sure that the burghers of Pontresina felt that it's good to be wanted.

When I arrived the Production Office was already swinging along in top gear, with Leonard Gmur at the wheel, but the whole film unit wouldn't be up to strength for another two weeks. Various members of the safety brigade were erratically winding their way across Europe frequenting various hostelries *en route*.

The tourist season hadn't begun and some of the hotels were not available, with the result that most of us had temporary accommodation at the Steinbock, a small hotel on the main street of Pontresina. A couple of days later Sean Connery arrived. I had not met him before and I wouldn't have known what he looked like had it not been for photographs Peter Beale gave me when we were trying to size up the Mafia for a suitable double. I'm probably one of the few people who has never seen a Bond film, though I'm well acquainted with Ian Fleming's family as they have a shooting estate adjoining Glencoe.

I suppose it was inevitable that I should get on well with Sean. We were both exiled Scots (though I only temporarily) and we both shared a common dislike for Switzerland; Sean's was more deep-rooted. We both spoke the 'Arr ye awright, Jimmy?' language, and I knew he'd get on with the Mafia.

"I only hope my knee's going to stand up to this proposed mountain-bashing, Hamish. I've had several operations on it."

"Best thing is to take it easy for a bit," I suggested. "A few gentle walks will break you in slowly." Ironically, his first 'gentle' walk almost broke him up completely.

Betsy Brantley and her double, stunt-girl Wendy Leach, arrived together. Both looked fit and no doubt were wondering what they had let themselves in for. Well, the following day they were let into a training walk with Sean. I sent the old guide, Alfred Schmidhauser, with the three of them thinking, quite wrongly, that he would take them on a gentle constitutional, as suggested. In fact, the route he chose would normally have been fine, but he had forgotten about the last of the winter snow which obliterated large sections of the path, with the result that it was thigh-deep in places (for Sean at least, as he is heavier). Sean's knee was particularly vulnerable and he hobbled back to Pontresina with his opinion of the Swiss substantiated. I felt a bit guilty about this disastrous walk because the other groups with Swiss guides fared better and seemed to have had an enjoyable day. This part of the crew training was in the hands of the local climbing school, though it came under my general umbrella.

The following night Betsy and I had dinner together in an adjoining hotel. After a couple of bottles of Dôle, Switzerland didn't seem such a bad place — that was till we arrived at 10 p.m. at the front door of the Steinbock Hotel. It was locked and as secure as the vaults of the Zurich Gnomes. Repeated hammering on the panels only produced echoes. Even Betsy, never known to procrastinate in a trying situation, failed to elicit attention by deafening and persistent shouting to the accompaniment of a drum beat on the door.

Not wishing to resign ourselves to a bivouac in the car or on the hotel steps, I prowled round the side looking for an open window. Lights blazed in the lounge and inside was a Swiss couple playing draughts. I knew they were Swiss as the man wore a 'William Tell' jacket and *lederhosen*, and the woman had her hair in a tight-fisted bun, no make-up, and a right-up-to-the-neck, no-frigging-about black dress. I hammered excitedly on the plate glass window and shouted to draw their attention, but they kept their eyes firmly fixed on the board like absorbed watchmakers. Unless they were stone-deaf, they must have heard our commotion — Sean admitted the next day that it awakened him two floors up.

Assuming now that they could possibly be totally deaf, I waved a white handkerchief, like a shipwrecked mariner who had spotted a ship after months of waiting, and mimed opening an imaginary door. I saw the woman cast a fleeting glance at me, only to return her guilty eye back to the board. I gave up and returned to the indefatigable Betsy.

"There's a couple of deaf mutes in there," I pointed up towards the lounge, "playing draughts and hard to get. Perhaps you'd better try, my bearded countenance is probably not the right image at this time of night."

Betsy was back in a couple of minutes, beaming.

"They're going to open the door. It pays to be a woman sometimes!"

Lambert Wilson arrived and was immediately sent off with two guides to continue his climbing training. Everyone who could walk, or who perhaps thought they couldn't, was pressed into training. To some, certainly in the beginning, it was akin to *bastinado*, but they were soon feeling the benefit of fresh air and exercise. Fred Zinnemann had his own personal guide, Hans Wohrle, a quiet, dark, bearded man who had little to say, but who looked after the 'gaffer' as if his very life depended on it. Fred required the least persuasion of all to take to the hills. I think he felt he was 50 years of mountain enjoyment in arrears.

Most of my days were taken up in helicopter recce work with Andrea Florineth, or in talking to Mr Zinnemann about possible locations. The Mafia swung into action, too, as the training graduated from paths and slopes to rock and ice. Soon there was an almost competitive element amongst the crew, and Peter Beale, among others, was ever striving to do harder routes. Talk of 'P.A.s' (special lightweight climbing boots) and 'front pointing' in crampons was the *lingua franca* of the crew members.

Norman Dyrenfurth had now forged his Second Unit into shape with Mo Anthoine and Tony Riley included in his film team, though he could draw on the Mafia and guides as and when required. No filming had started yet and it wasn't clear when it would. Fred Zinnemann held the cards and he was holding them close to his chest. Both Wendy and

Betsy proved good natural climbers and it was a relief to know we would have no problem in that quarter. Lambert was coming along quite well too, though he did have trouble with the mechanics of rope work. Sean, on the other hand, with two years in the Royal Navy behind him, knew his ropes — and his knots — and he could handle belaying with conviction. However, the two male stars didn't fare so well with their nailed boots, the standard footwear of the period. But even old hands such as Martin, Paul Nunn and Ian Nicholson had trouble with these when doing their doubling. Climbing in nailed boots involves a different technique where the boot, or rather an individual nail, has to be placed on a rugosity with the precision of an ophthalmic surgeon — then you have to have the nerve to stand up on it.

Zinnemann felt that Betsy's accent required refining to that of an educated Scots girl, even though there was no trace of an American drawl in her everyday conversation. She has the enviable ability for an actress of being able to adopt the accent and nuances of any given place.

Part of this training was elocution and she had to tape passages from books or plays which were then played for Zinnemann. One day I received in the mail a paper from some Cambridge undergraduate who had read my book *Climb to the Lost World*, a saga of climbing in the slime forests of Guyana. He sent me an article on the habits of the orifice fish, found in that region. I suggested to Betsy that the habits of this creature would make a good voice-lesson tape for her to read. Fred Zinnemann didn't find it as amusing.

By this time we had been diluted into various hotels. The Mafia, who were segregated in a hotel apart from the rest — probably a company policy to keep all bad eggs in one basket — were dissatisfied. As their carousing extended far into the night, they suffered the same fate which befell Betsy and me, that of being locked out. Their subsequent complaints to Leonard Gmur, the Production Office Oberführer who had arranged all the accommodation, met with stony silence.

I had moved to the nearby Hotel Walther, a palatial establishment, which also now catered for Mr Zinnemann and other members of the upper echelon. I think I must have been very much at the lower end of this bracket, for my room wasn't big enough to swing a Manx cat in. However, the hotel's imposing façade was fancy enough to put off my Austrian baker friend, Arnold Larcher, when he called. He was over in the Engadine to take a client up the Biancograt, but was so overawed when he entered the foyer that he turned tail and fled. I did meet him the following week when he returned. This was to be the last time I saw him.

Most of the Mafia were now in residence. Paul Nunn had been with Tony Riley on a winter attempt of Mount Everest's West Ridge and both were recovering from hepatatis. Others had been on various expeditions since I last saw them, so there was much to talk about.

Whilst we were busy with our daily ploys, Willie Holt, the Art Director, was constructing sets. Willie is a fascinating man, tall and lithe, with a shock of dignified grey hair and matching moustache. I think he would have made a good lama or an archbishop, for he radiates a serenity; the sort of person Fred Zinnemann might have cast as the Good Samaritan. Willie and his dog 'Maquis' were great favourites. Man and dog were inseparable and, as often happens, Maquis held an uncanny resemblance to his master. Willie accepted all mankind on an equal basis . . . there was no bullshit with him. As an Auschwitz survivor, he still had a branded number on his arm, but would never have dreamt of showing it to anyone. Here in Switzerland, Willie had the daunting task of building a three-storey hotel, or at least enough to make it a convincing set, high in the Roseg Valley. This valley is a National Park and therefore had severe restrictions. The building had to be constructed within a specific time period, used for 'x' days and dismantled, leaving not a scrap of evidence that it was ever there. No motor vehicles are allowed in the valley, so the horse 'taxis' did a booming trade. But I must give credit to the Swiss, as they were most co-operative once these rules had been adhered to and gave special permission for helicopters to fly to the site when required.

Usually my day was spent in flying and landing on glaciers and summits with Ueli or Capito. Mostly I was with Andrea Florineth on these trips, or Norman, Pepino and Fred Zinnemann. Meanwhile the crew was continuing with their mountain square-bashing. The most intensive climbing schedule involved the actors and Lambert must have felt a lot of pressure. The local guides knew he was to act the Swiss guide and they were sceptical, to say the least, of his mountaineering capabilities, with the result that he dreaded going out with them. It was Martin Boysen, his double, who took him under his wing and I believe that, had it not been for Martin's patience and expertise, Lambert would have had a much rougher time.

Generally speaking, things were panning out with the local guides. Tony Spinash (whom we called Tony Spinach), for example, was one of the older guides — a sort of continental version of Popeye with an eye screwed up in a perpetual wink. Like many of the guides, he wore that other badge of office, a flat white cap, and like a hyper-active kid, he was always on the move, always wanting to lend a hand.

Tut Braithwaite, an old friend of mine and a member of the Mafia, had a great camaraderie with Tony. He would hold the guide by the shoulder and say, with an air of great seriousness, "See that patch of scree over there?" He pointed and repeated slowly, for Tony's English wasn't too good. "I want all that turned over — you've got fifteen minutes, Tony — get to it."

Tut's particular brand of humour was often quick on the draw. Another time, he was in a local sports shop, having one of the crew fitted out. An assistant approached. "Are you a guide?" she asked. (Tut wears a white cap.)

"No," Tut replied, dead-pan, "I'm a climber."

Incidentally, Tut has done virtually all the major routes in the Alps.

Even 'Alfred the Great', as I called Alfred Schmidhauser, had an endearing personality once he understood the set-up! The guides met me each morning at the Production Office to get details of the day's activities. One day, Alfred was climbing with two of the crew on training rocks close to the Morteratsch Glacier when Joe Brown, who was leading Betsy up a pitch, arrived at a ledge close by and put the climbing rope round a tree as a belay. For some inexplicable reason Alfred felt this was an act of extreme vandalism, which would have won him a distinguished conduct medal in an environmental group. He started bawling at Joe. But Joe, the master-plumber, was always the paragon of patience and virtue, and kept his cool. He decided in the interest of Anglo-Swiss relations to belay elsewhere.

Crew training and location-finding were gobbling up time and Peter Beale was getting worried as to when shooting would start. There was all sorts of speculation. I could sympathize with Fred Zinnemann, for I feel the same way when about to start one of my mini-projects. It's knowing that once you start, that's it. In this case the catalyst which started things rolling was the winching training. Everyone had done helicopter winching now except Fred and it was arranged one afternoon that I would go with him and Ueli the pilot so that he could be shown the basic technique. We did this high on a lonely alp above the Morteratsch and though Fred was, as everyone else, somewhat jittery, he was like a dog with two tails once it was done — it seemed to instil a new confidence.

Peter Beale phoned me up later that day.

"Hamish, I owe you a bottle of champagne, you've done it. Mr Zinnemann said the film's about to start. He made the decision after the winch training."

"I don't think it's anything to do with me, but I'm glad that we can get cracking. I had a feeling that we might be celebrating Hogmanay in one of the crevasses of the Morteratsch Glacier."

In some ways, being involved with a large film crew reminded me of being back in the Army or on an Everest expedition. All looked to the 'CO' or the 'RSM' (Peter Beale) for guidance. Mr Zinnemann, as Commander of this company, had widespread respect. In his quiet way he got things done and always had time to say a few words to everyone.

As 'Zinnemann' happens to be quite a mouthful (especially in Hollywood, land of the nickname), it was most often reduced to 'Mr Z.' (pronounced 'zee' over there). One did not address him as 'Fred' unless given specific permission, as he is quite a stickler for discipline in the ranks. There was a further nickname coined for Mr Zee in Hollywood — 'The Iron Butterfly'. The meaning became clear to me later. I was blissfully unaware of this protocol and had got used to calling him 'Fred'. Indeed, all the climbers followed suit. Mountaineering is a great leveller of society; even the most exalted in the climbing world are known by first names.

Though the Mafia got away with their nocturnal carousing, not so Betsy. She soon realized that she was expected to be Kate, a demure, well-brought-up girl with impeccable manners, 24 hours a day. Betsy, full of the freedom and warmth of the South, found this quite alien to her extrovert nature when off the set. Criticisms of our meeting with the Mafia 'in public' were channelled through me with the result that we had to meet up with the boys in lesser-known hostelries. Andrea was helpful here, for he knew all the restaurants in the district and these, in turn, were injected with evening revelry.

Although the weather was bad in most other areas of the Alps, it was fine in the Bernina. It was so good, in fact, that by the end of June the lower part of the Morteratsch Glacier, where Fred hoped to shoot some of the larger scenes, was more like a deserted building site. Dirty rocks littered the ice which resembled slag and I'm sure that, had we used it for the film, it would have looked more like the aftermath of the Aberfan disaster.

We inspected it one day with John Richardson, who was in charge of special effects; as a last resort he suggested that he could try high-pressure ablution with hoses to cleanse about half a square mile! Other ideas, such as putting salt on the ice, were not permitted as this affected both livestock and flora further down the valley. We were told that the use of salt on the roads was discontinued for this very reason. Polystyrene or other forms of camouflage were out as well under the local anti-litter laws. These were very stringent, and rightly so. Later, we even had to carry our portable loo 'piggy-banks' back to the village sewer for disposal.

John Richardson didn't fare as well as Hercules mucking out the Morteratsch Stables as the high-pressure pump filters blocked, owing to sediment in the glacier water; anyway, the task was akin to scrubbing a ploughed field with a toothbrush.

We had, in fact, more casualties in post-prandial activities than during working hours. Paul Nunn literally fell victim in this hazardous crepuscular period.

A party which had started like a film in slow motion in the cutting room was later injected with vigour when it fragmented and re-formed in the cocktail bar of the Hotel Walther. This is the palatial hotel into which I was billeted earlier and the place where Arnold the guide turned tail and fled at its sheer opulence. The Mafia, as they knocked back a constant stream of free drink supplied by Peter Beale, were unaffected by the presence of the establishment. As a matter of fact there were complaints that the men's loo was too distant. Joe and Mo, observing that the balcony was a mere fifteen feet above the ground, decided to take a short-cut and jumped, breaking their fall on a pile of garden debris. After they relieved themselves they promptly climbed up the wall and merged back into the throng. This sequence was repeated by other members of the Mafia who all prefer to climb rather than walk. Paul Nunn, with a whoop, also launched himself into space and, like a rather erratic boomerang, returned to the action. His downfall came when he tried to repeat it, after further imbibing. This time he missed the 'springboard' of garden waste and all fourteen stone of him came down heavily on the flagged garden path. He folded like a clapper board and his knees made contact with his chin, effectively splitting it. However, as it was time to go anyhow, they all retreated to their doss, the Cresta Hotel, and Paul repaired the aperture with tape to stem the bleeding. He then retired to bed, an enormous expanse of elaborate quiltwork and carved wood. He had just settled when the phone rang. He picked it up. It was his wife. Paul mumbled into the mouthpiece. His speech was impaired, but he didn't want to tell his wife, Hilary, what had happened.

"You're drunk," she accused. "When are you coming home . . . ?" No sooner had this encounter finished than Paul had another crises on his hands. Speaking into the phone had disturbed his wound and now the once pristine white sheets were a mass of blood, "just like the great chainsaw massacre", he said later.

He phoned Rab Carrington and Rab came down from his room to help stem the bleeding. A barrage of heavy sticking plaster eventually did the trick and Paul subsided into a fitful sleep in somewhat messy bedcovers. There was no time to have the wound seen to in the morning, for he was scheduled to take Sean out for snow and ice training with one of the guides. The location was high on the edge of the Pers glacier and the first lesson was braking with the ice axe. This is a method of stopping in an emergency. With Sean safe on a nearby ledge, Paul launched himself boldly down a slope and, trying to instil confidence in Sean, inserted the pick of his axe in the approved manner. However, the procedure disturbed the large bulging mass of plaster which hung from his chin like an udder. It was in fact now a reservoir of blood. This ruptured and splattered over the snow cover more effectively than any special effects. It was also somewhat off-putting for his star pupil. However, as Paul said later, "Sean already thought me and the rest of us mountaineers idiots." It wasn't a good day for Paul, or for Sean for that matter, for Sean wrenched his bad knee when he slipped into a hole on the edge of the ice.

We had a Work Schedule which was printed by the Production Office each day and in this, after the balcony scene, I jokingly made a passing remark to the effect that 'members of the Mafia should refrain from climbing after working hours'. Fred Zee, who read this and had also seen Paul's grizzly chin (though by this time it had been properly stitched), instigated an enquiry. This was presided over by Peter Beale, who had in fact been the perpetrator of the re-formed party. The other members were Paul and myself. We tried to keep straight faces as we went through the farcical formalities in Peter's office. We were three guilty men putting ourselves on trial.

Andrea and I now had another task — to find a new glacier which would have that fresh Monday look, yet have few dangerous crevasses in which to lose crew members. Meanwhile, the search for 'the' big crevasse which was to be used for the discovery of the thirty-year-old corpse still hadn't been resolved. Mr Zinnemann knew exactly what he wanted and he set his sights high, or in the case of the crevasse, deep. At last we located one which was within possible walking distance from the top of the Diavolezza cable car in an emergency (though helicopters would have to be used to commute). The crevasse was about 200 feet deep, with a shelf some 130 feet down and not too desperate to get into. Access was via an ice arch, the width of a train tunnel. We took Mr Zee, Pepino and several of the other executives down and it passed its audition.

I realized that it was going to take about two weeks to prepare it, for platforms were required inside and ladders had to be fixed for access. Above, a camp had to be erected with landing pads for the helicopters and the whole area probed for covered crevasses. Also, several tons of equipment had to be shuttled to the site. I took Mo off the Second Unit as he is a qualified scaffolder and, with Rab Carrington and Big Ian, a start was made on this major location.

Back at the Mafia ranch there were continuing rumblings of discontent over their 'top security' hotel. After Paul Braithwaite, who wasn't really a night owl, found himself locked out of his hotel twice at the early hour of 11 p.m. they finally decided to move out. I had already evacuated my room as Andrea had found me a flat in St Moritz. Sean had moved into a flat in the block next door. Rab Carrington and the boys also flitted to a flat in St Moritz, where exhibits of delicate china and brick-à-brac were in turn moved to more secure, remote cupboards to avoid demolition. Paul Nunn had two crucifixes in his bedroom — much to his disgust. As there was only one double bed, it was agreed that this should be occupied on a rota basis as and when wives and girlfriends came to visit.

At the same time as Mo, Ian and Rab were working on their building project in the crevasse, Willie Holt had started constructing a mountain hut on the moraine below the top station of the Diavolezza cable car. This was to be a full-sized mountain hut in the old style, with a simulated stone balcony. It was an enormous task and all the building material had to be flown up by helicopter.

By this time, the sad-eyed, restless cows had gone up to the high pastures for their summer holidays. With most of the winter's snow melted off the lower Alps, crocus and gentians raised their heads to a new world. The Engadine is particularly well-endowed with wild flora and it is a delight to see the advent of spring. In fact, almost without our noticing, spring had slipped into summer and whilst other parts of the Alps were deep in snow, we were basking in sunshine.

It wasn't until we had various towers and ladders fixed in the crevasse that the effects of the heat wave became obvious. It was playing havoc with the glacier. The crevasse was opening its jaws like a giant leviathan in a slow motion yawn. Scaffolding was starting to list and the ice tunnel leading to the working area, 'the platform of Jonah', as I called it, was bombarded by falling pieces of ice and the perpetual spray of melting water. It became so dangerous that we even contemplated working in the crevasse at night by floodlight when the temperature dropped. Also, up here at 10,000 feet, the glacier now looked as if it could do with a face lift. The surface resembled the pallor of a corpse that died of smallpox.

Though the unit had been accident-free so far (except for one minor mishap), the same couldn't be said for other mountaineers enjoying the peaks at this period. There were a number of serious accidents and almost every week someone was killed. I was on several rescue missions with the helicopter crews as I was flying so much with them.

During this period the Mafia lived life to the full. There were long, though not taxing, days on location, but the Mafia visitations to the wine houses left them with a sleep deficit and a high level of blood alcohol. Nevertheless, this didn't detract from their climbing and each weekend they put up hard new routes. Some of them even drove the 400 miles to Chamonix for a day. Mo thrives on such undertakings, having motored to the Himalayas several times. One such weekend, he took Joe's car; Joe was left with Mo's van. Both vehicles ended up off the road and both vehicles bent: Mo by trying to take a corner flat out, and Joe in dispute with a fence designed to deter chic Swiss goats.

Sean had made arrangements to go to Scotland for a few days to take part in a golf tournament and to do publicity on his previous film. Before his departure I had found that I required reinforcements for the Mafia as both First and Second Units were working full steam ahead. Also, the guides were now in demand as it was the peak tourist season.

I phoned through to Edinburgh to recruit Murray Hamilton, one of Scotland's leading climbers, who had worked with us on the Glencoe Outside Broadcast. He agreed to come out right away, but asked if he could return to Scotland for one day the following week — to get married! Being used to the dedication of climbers to their sport, or anything connected with it, I was not too surprised at this lightning wedding arrangement, but I suggested, "Why don't you bring your future missus out here with you? You will probably be able to get a flat; certainly a hotel. I've another idea, too, but we'll talk about it when you come out."

The idea I had in mind was to ask Sean if he could give Murray and his wife a lift back from Edinburgh in his private charter plane. Sean agreed to this, and in due course Murray arrived in Pontresina with his new bride after a champagne-punctuated flight.

Through some planning blunder, it had been decided that the crevasse scene (when the guide discovers the deep-frozen body) should be shot in midsummer. Working on top of the glacier during the day was sweltering and in the depths of the crevasse it was like an upgraded Venetian canal. Our particular crevasse was now dripping and wilting dangerously. Mo, who had been working in it like a latter-day Charon, suggested that I should inspect it again. "It's pissing wet, with snots of loose ice," he stated. It was, and as soon as I saw this, I realized that we would have to salvage whatever equipment we could and get the hell out. All the work of the crevasse squad was now written off and I looked at a neighbouring crevasse which was even deeper. When we first came to the site, it had been a mere 'slitling', some five feet wide. Now it gaped 30 feet across. I took Mr Zee in tow to check it out. He liked it, so did Pepino. Now it was all hands on deck or below 'deck'; the scaffolding of the old crevasse was dismembered and ladders dismantled, all without incident, fortunately. We knew that shooting must start in our new slot as soon as we had completed the work; four days later the cameras rolled on a platform 100 feet beneath the surface of the ice.

That evening when Betsy and I were having a late meal, the phone rang. It was Hans Spielmann from Austria. Arnold Larcher, the guide, had been killed, struck by lightning in the Zillertal Alps. The funeral was to be the following day. I realized how upset Hans was; he was desperately trying to contain himself. I had known Arnold for many years, but I didn't live beside him as Hans did. He and Hans were constant companions on climbs and ski tours.

"What time is the funeral, Hans?"

"Nine o'clock."

"Right, I'll be over."

I was deeply impressed by the funeral ceremony. Arnold's death had made a big impact with the media; there was even a film unit there. He had been scheduled to leave on the Makalu Expedition the day after the funeral. Besides, his wife had given birth at almost the exact moment Arnold was killed. Hans spoke to the doctor who had examined him and was told that there wasn't a mark on his body, but that all the points of his crampons were bent at right angles by the lightning.

I went to the churchyard with Hans. I was dressed in my climbing clothes like most of the other mourners. As Hans and I approached the church, a shaft of sunlight caught the gold of the onion-topped spire and, above, the prodigious backdrop of the Wetterstein hung over the scene — one of the most impressive mountain walls in the Alps. My eye automatically traced a new route pioneered by Arnold some years previously, and to its right a new climb I had done when I lived in this warm-hearted Tyrolean village. As Hans was directly involved in the service, the Makalu Expedition members and I took our places with the local rescue team. The church and graveyard were overflowing; there were mountain guides from all over the Tyrol; gendarmerie and helicopter crews.

The coffin was then borne in procession round the village, as is the local custom. When the circuit was completed we returned to the churchyard. Amazingly the tail-end of the procession was only just starting off. Friends of Arnold, the priest and leaders of rescue groups gave short speeches at the graveside, outlining Arnold's achievements; simple statements sincere in their expression and most touching in that rugged mountainous setting. It seemed somehow to symbolize the transitory passage of man and the permanence of those lofty limestone peaks.

It was late when I got back to my flat in St Moritz, but I gave Joe a ring.

"How did it go, Joe?"

"I know what you're going to say, or at least think, Hamish," were his opening words. "Anyway, I'd better tell you. We were working down on the bottom platform when we had a bit of an accident. A lump of ice fell off the wall of the crevasse. It smashed a Panavision camera, knocked a lens off and then hit Fred. But he's all right. As a matter of fact, he caught the lens as it was sheared off the camera."

"Where was he hit?"

"On the back of the neck, but he's okay."

"Well, I'm damned," I breathed, thankful that no one had been hurt. "It would happen to Mr Zee, wouldn't it?"

Our crevasse camp was one of the major attractions of this particular travelling circus. It looked like a builder's yard with scaffolding, planks and alloy structures. Initially, the helicopters landed on a thin catwalk of ice between two huge crevasses, with the rotorblade overshadowing each chasm. Getting in and out could be hazardous, so we moved the landing site to the open névé nearby and built several bridges to span hidden crevasses. The catering tents then occupied the old landing site. So close were these tents to the crevasse lip that the walls of these Agincourt structures on the upper side formed an extension of the lower crevasse wall. The crew was wary not to lean back on their chairs on that side of the tent, especially after a gourmet lunch of lobster or smoked salmon. However, it wasn't long before they got used to working on the glacier. Ironically, that was the most dangerous time for us. We had to play watch dog constantly. Otherwise, people would wander off and be quite blasé on terrain which a few weeks previously would have scared the smoked salmon out of them. It was the old problem of a little knowledge being an extremely dangerous thing.

For a jump scene, where the three actors had to leap a crevasse, we required a snow projection on either side as it was too wide to jump otherwise. We manufactured these by inserting logs into the glacier surface so that they projected part way over the void, then 'plastered' each structure with snow which adhered to the chicken wire we had stapled to the wood. The plastering job was delicate and exposed, for Joe and Tut Braithwaite had to hang free on harnesses whilst carrying out this operation.

As Tut commented when suspended over a 200-foot drop, "It's not the sort of job you should do if you've loose bowels."

Peter MacPherson was an assistant cameraman on the Second Unit and as such he had to be a competent climber. Months before, Peter Beale had given me a ring, saying that MacPherson was applying. It transpired that Peter MacPherson had been on my snow and ice course in Glencoe years before.

Occasionally, Peter, Herbert Raditschnig and Arthur Wooster had to help out with the main unit when several cameras were required. One day, when the glacier was snuggling under an impossible smothering of snow, a scene was being shot at a railway station. This depicted Kate, after the long journey from Scotland, snoozing on the station bench whilst Dr Meredith paced up and down awaiting the carriage which was to take them to their hotel. Peter describes what happened that day:

"I was there with Herbie, armed with a battery of cameras. Herbie was determined to get in on the action and soon had his chance when Mr Zee called for a second camera on a shot that was set up and ready to roll. After a withering look from the 'Iron Butterfly', Herbie and I settled down to a mundane day of frantic shooting of cutaways. I took the mag off and put it in the back of a Unit vehicle in view of two of the 'Italian mob' (Pepino's camera crew). When we got back to the hotel that evening, Herbie asked me, 'Where's the mag?'

"I was stunned. Rushing outside. I leapt into a car and hared back through the valley and up to the station. Beyond Pontresina I was almost at the small station when I met the 'sparks' wagon. It was a narrow, single-track road and I had to brake hard to avoid a collision.

"'Got your ticket home, mate?' a cheery voice called out from the cab. Beside the driver on the seat was the missing magazine.

"Apparently, a local had seen me putting the magazine in the van, had picked it up and made off with it through the crowd of extras. However, the hotel manager spotted him and closed in before he completed his getaway. It turned out that the thief was going to hold the Ladd Company to ransom with this film, obviously unaware of the 'mettle' in the Iron Butterfly."

It's inevitable that tensions build up in such a costly operation as a feature film, with Hollywood moguls breathing down one's neck. But I honestly don't think that Fred Zinnemann was affected by such pressures. To him, the film was all, and he became incredibly involved with every aspect of it. No detail was too minuscule to be ignored. Betsy and Lambert, as newcomers to the screen, and in leading rôles, bore at least part of the brunt of this microscopic scrutiny.

One afternoon Fred called me over to the old ice rink at the nearby village of Samaden, where a plan of the crevasse was chalked out on the floor. This was for rehearsal purposes, to save valuable shooting time when on the glacier. The Iron Butterfly was obviously in a stern mood. Things hadn't been going right. I could tell from Tony Waye's expression that sparks had been flying. Tony had confessed to me earlier that he sometimes felt more like a referee than Assistant Director.

When I entered the long stark dusty building, Fred was in the middle of the large floor, gazing at the chalk marks with a faraway expression. In his hand was the now well-worn script.

"Hello, Fred."

"Oh, Hamish." He spoke in *sotto voce* tones as if about to divulge a secret formula and, grasping my elbow, walked me slowly towards the end wall of the building. "Hamish, Betsy needs to do more voice work. I know it's her first movie, but I'm worried."

I didn't say anything, for I sensed what was afoot. There had been persistent inroads into Betsy's free time until now she was working from 5 a.m. until 6 or 7 p.m., and doing voice work for an hour after that. I knew that she was about at the end of her tether.

"Well, what's on your mind?" I was somewhat abrupt, for, though a mere amateur, I didn't agree with his assessment of Betsy's English. For the past few days her scenes had been straightforward, and she had the ability to pick up a required accent in a couple of days. Her accent was so natural that many of my friends didn't even know she was American. Even so, she was working with a voice coach all week and on Saturdays.

"I want to suggest, Hamish, that after her voice lessons at six, she has dinner at her hotel, instead of at your flat; then does a further hour after her meal, and also on Sundays."

I had considerable trouble in controlling myself and after counting up to five, I replied tersely: "Fred, if Betsy works any more hours than at present, she'll be a mental wreck, and you know it. She's not a robot!"

He was obviously taken aback by my vehemence and, slightly fluttering, he returned, "It was just a suggestion, Hamish, something for you to think about. I'm not advocating it's done right away, just consider it and let me know."

The matter was never raised again and we continued to be the best of friends. I liked Fred despite his iron rigidity; one had to admire the way he got what he wanted. In fact the whole gambit of Betsy's working longer hours may have been a psychological ploy on his part to keep her on her toes. I didn't tell her of the conversation.

Pressure was taken off Betsy soon enough, for within the next few days all the attention was focused on Lambert, or rather on Lambert's pimple. In real life, a pimple is a non-event, with the exception of discomfort to the owner for a few days. However, in film language it spelt disaster by interrupting continuity. Probably if it had been allowed to run its normal course it would not have developed into Lambert's *bête noire*, but he got so upset by everyone making a mountain out of his molehill that matters worsened.

After two days, shooting ground to a halt. Even some of the interiors which were to be shot in London were now being prepared in a newly made indoor set in a nearby village. A massive insurance claim was submitted for the loss of shooting time and assessors flew over to Switzerland to see the pimple for themselves. By this time it had matured into a boil of generous proportions which no amount of make-up could camouflage.

In the mountains we had time to catch up with other preparatory work and the Second Unit under Norman Dyrenfurth was in full cry, but here, too, it wasn't all a bed of roses. A conflict of personalities had developed between Arthur Wooster and Norman. At the same time, a bond had formed between Herbie Raditschnig and Norman, both of whom have a mountaineering background. Arthur, a superb camera-man, has very definite views on cinematography but, not being a climber, his suggestions were not always feasible, even though his instincts for composition were. It was unfortunate that this divergence of views occurred, for it made life difficult for the rest of us. Arthur didn't lack courage — 'even in the cannon's mouth' — and he asked to shoot in places which were positively dangerous. He was later given a BAFTA award for his work on this and other films.

Further friction was caused by the safety precautions. Though the Mafia were some of the most experienced climbers in the world, they never neglected people's well-being. In fact they often brought to my attention things that were not as they should be. Usually this was in connection with the Swiss guides and dealing with such problems was a precarious tightrope for me to tread. Though the lads didn't have a great deal of respect for the older generation of Swiss guides, they made a valiant attempt not to rock the boat.

During the latter part of the film, when the guide climbs with Dr Meredith, and Kate is left twiddling her thumbs in a hut, we set the scene on the north face of the Porta da Roseg, a very steep ice wall, severed at the base by an enormous bergschrund. The wall led on to a narrow col, very sharp and providing another 'not to fall down' drop to the glacier on the south. Access would be out of the question except by helicopter. I asked Ueli, our chief pilot, how he felt about landing or winching us on to this col.

"Yes, Hamish," he replied in his deliberate, guttural way, the mountaineer in him responding to the challenge, "let's try."

We piled into the helicopter after lunch. As we approached this hostile place, I grew apprehensive upon seeing the gunmetal ice, lethal-looking and forbidding. The slope at the bergschrund at the base of the wall looked like a grubby throat slashed by a razor.

There wasn't much room to take evasive action in here. Helicopter pilots like a bit of elbow room, space to slip down sideways in the event of sudden downdraughts. This narrow col could be a death trap except for the most able pilot. Ueli coaxed the machine towards the crest as if it was a shy horse and gently set part of the skid on to the cornice. It was an uninterrupted 1,000 foot drop to the Tschierva glacier.

He gave a nod as his voice came over the intercom. "Yes, I think we can do it all right without winching, Hamish."

"We'll cut out a ledge for you to get a full skid on, Ueli."

"Okay, is that all?"

"Yes, we'll come back tomorrow and go down the face."

Early next morning the mountains were still snoozing under an overnight frost, then the leggy rays of the sun came striding through the crazy seracs on the glacier. Ueli took the rescue Alouette warily into the confines of the Porta. Joe Brown and Andrea were in the back.

"Okay, Hamish, one at a time." Ueli's voice came over the headphones.

"Thanks, Ueli, we'll give you a call on the radio when we want to get back to the bar — we'll need a drink."

Gingerly I stepped out, testing the snow. It seemed fine and I quickly moved back towards the root of the cornice and motioned the others to follow. Andrea stepped out, followed by Joe. However, with the loss of passengers, the machine rose slightly at the precise instant that Joe's harness caught on a projection by the cabin door. The result was that he fell forward. He was safe enough, for I had a grip of his harness and in any case he was some four feet from the drop. However, as he touched down (in a rather undignified heap for such a famous figure), the helicopter skid, gently as before, came down on top of him. It reminded me of that ancient Indian custom in which prisoners were executed under the foot of an elephant.

As his harness was now free, I gave Ueli the thumbs-up sign for take-off and Joe was instantly relieved of pressure on his lumbar region. He was unhurt.

The ice wall was just what we wanted and it was later used for a major sequence which involved Martin Boysen doubling for Lambert and Ian Nicholson standing in for Paul Nunn, Sean's double, as Paul had to return briefly to Sheffield Polytechnic to supervise exams. We dropped 700 feet of rope down the face and cut perches for the camera crews at regular intervals, whilst the doubles climbed a parallel line. It was a remarkable and dangerous climb as the ice was ivory hard and the belays minimal. By the time they reached the col, Martin, who had been leading throughout, was in a vile temper. He had been step-cutting for six hours and feature films became his whipping boy. Fragmented ice like disintegrating chandeliers cascaded down on Ian, 80 feet below. "Bloody films; six hours' hard labour and mucking about and the bloody thing will never be used anyhow!"

On another hazardous location, Capito was our driver. This time it was a mission to the North Ridge of Piz Palu, a pillar with the edge of a blunt cleaver which buttresses the face of the mountain in an elegant sweep of ice. Norman had various cameras set up on the two summits of the mountain. Rather than the doubles climbing the East Ridge from the glacier, we proposed to fly them to a point half-way up. We had to do this twice, once with Herbie to get close-ups and then with the doubles on their own.

I went aboard with Capito to size up the possibilities. It was a superb day and I felt envious of such weather. In Scotland, when helicopter flying, we are always pummelled by high winds. Capito's style was the antithesis of that of Ueli. Capito was more of a rally driver — stomach-knotting dives and gripping manoeuvres. He was, however, a talented pilot. With no sign of nerves, he took the Lama, a small, powerful helicopter which can lift its own weight, right up to the edge. (The Lama helicopter also has the world's high altitude record of almost 41,000 feet.) Below was a rock obtrusion.

"How about that rock for resting the skid on, Capito?" I pointed through the Perspex floor. "Or is it a winching job?"

"I think that I may be able to put down there, Hamish. How's the rotor?"

"You have about two metres between it and the ice."

"Umph," was all he said. He gently brushed the rock with the skid; all the time I was gazing mesmerized at the rotor and the blue-green ice just beneath it. "*Ja*, we can do without winching. Can you tell the climbers that they'll have to get out not clumsily?"

"Don't worry. It's our skin, too, Capito, and it's a long drop to the glacier!"

Towards the end of the film, Capito, like the rest of us, must have been feeling the strain of long hours and dicey situations. One day when I had to get everyone down from a high-glacier location pronto, with an electric storm flashing towards us, Capito was shuttling the 60 crew members back to base as fast as his egg beater would fly, as well as picking up underslung loads. Paul Nunn and I were crouched in the approved head-retaining posture, down on one knee, hand over cap and eyes to prevent snow pellets from blinding us. I felt something cold brush the back of my hand, like the touch of death; it was the tail rotor guard. The propeller was inches from my head and only a foot from Paul's. Capito had apparently forgotten about us when he swung round to pick up the underslung load. Both Paul and I felt distinctly drained by this incident. That Old Man with his scythe doesn't come much closer. I also had the macabre background knowledge of someone back home who had walked into a tail rotor. He had ended up in the coffin like a pile of spare parts.

There is a summit scene in the film where the guide confronts the doctor, rather unrealistically I felt, telling him that he, as a married man, was stealing a young girl's life away. Fred Zinnemann wanted a snowy summit for this, pristine white and with a large drop on one side. It was more difficult finding such a place than we expected, and Andrea and I summit-hopped dozens of peaks in search of a suitable location. After selecting three or four we took Fred and Pepino round to view the pick of the tops. We had to be careful not to scare Fred, and I told Capito that aerobatics were out. So did Peter Beale, in no uncertain manner. But it was inevitable that at times we were caught in downdraughts or in mist or cloud and had to take evasive action. In the end, Fred decided on Pic Corvatsch for the summit scene location. This had the advantage of being close to a cable-car station, which terminated only 20 minutes' climb away. The snow was messy on the summit, which was considerably lower than the other peaks we had considered, and Capito suggested that we could use a rotovator to 'bury' the grotty snow.

"Not a bad idea, that." I slapped him on the shoulder; we were sitting beside the Alouette. "I think I'll plant some late cabbages too, because it looks like this film will never be finished!"

Next day, we must have presented a strange sight to passing climbers, crazy horticulturists ripping the surface off the summit snow with the large noisy rotovator.

One evening, after the day's shooting, I flew with Fred, Capito, Sean, Lambert and Tony Waye for a quick rehearsal on the summit: the confrontation between guide and client. Sean, who over the past month had found Lambert somewhat immature, was impatient with the Frenchman's inability to master ropework. The result was that, during the airy rehearsal, poor Lambert was shaken by Sean like a rabbit worried by a dog. This was not how Fred had envisaged the action of a virile young guide towards his rival in love, and he toned it down.

Probably the most dangerous aspect of making a film in high mountain country with inexperienced people is the problem of crevasses. A pristine expanse of snow, clean as a freshly laundered sheet, does not look exactly sinister to the unwary. The problem was magnified when Sean had to stagger across a wide glacier slope at a point where the guide has been killed and Sean, the doctor, has to go down to the mountain hut for help. Mr Zee didn't want his slope desecrated with boot prints, yet we knew there were numerous crevasses, any one of which could have meant a cold compress. Paul Nunn, Sean's double, had to do the most dangerous part of this stagger across the glacier, but Sean himself would have to enact the final approach to camera.

I had an idea and asked Capito if it was possible to probe the snow from a helicopter. That is, if we could hover a few inches above the snow cover.

"I know that these probes could have an argument with your main rotor, Capito; we certainly don't want that, but what if I stayed in the cabin behind you and have two people, one on either skid, with their probes coming in to the cabin towards me?"

"Ja." Capito was thoughtful.

"I could then bend them so that each would go out opposite doors, and miss the rotor blades."

Capito, who was standing beside the Lama, looked at me and then looked at the machine. "Ja," he muttered again slowly, rubbing his hand across his chin. "Ja, Hamish, I think it will work. Let's try it." That was the thing about Capito, there was no holding him back once he'd made up his mind.

Later that morning, Joe Brown, Garry the winchman, and myself climbed into the helicopter. At least Capito and I did; Joe and Garry secured themselves to the skids. The area we were going to probe was on the vast Pers glacier, which resides under the shadow of Piz Palu. With Mr Zee and Pepino I had already selected a slope for the sequence; it ambled gently uphill towards the central pillar of Piz Palu. It was, as Joe observed dryly, "as bare as a badger's ass." We landed briefly and assembled the probes. Once Joe and Garry were comfortable on the skids, they poked the ends of their probes into the cabin where I was sitting. I grasped them, one in either hand, holding them across the front of my body like a St Andrew's cross. Capito wound up the motor again and we rose a few feet, then moved forward.

"Okay, Capito, let's start here, straight up the slope."

"Fine, Hamish."

A few seconds later I gave Joe and Garry the thumbs-up sign and, as they inserted their probes, I felt the thin rods sliding through my hands. It seemed to work well as they angled up through the top of the doors, well clear of the main rotors.

"Hamish," Capito gave a laugh, "anyone seeing us will think the Lama is walking on very thin legs."

We made our funereal way up the long slope, putting in red slalom flags to mark the safe route. These were inserted every 50 metres or so. We intended to remove them just before Sean commenced his walk. Having found and marked several big crevasses with 'trapdoor' coverings, we landed back at our starting point; looking across the white expanse, the lonely red flags led up for about half a mile towards Piz Palu. They resembled drops of blood in that stark, sterile environment. We sat on the snow eating a late lunch in the scorching sun and then reluctantly piled into the Lama to start another task.

Imagine my dismay when, two days later, one of the guides told me that some descending *Hoch touristen*, seeing the providential line of flags leading back to civilization, followed them, effectively creating a deep trench in the soft snow which was quite irreparable. It obviously wasn't going to be possible to have a completely virgin snow slope for Sean's walk, so Fred and Pepino agreed to use the freshly made trail.

One of our most difficult tasks was to stage a stone fall, in which the guide is killed. Though the Special Effects brigade manufactured simulated rocks, there was no way that they could get into the climbing situations required for imitating the rock falls which Fred had envisaged. It was to be a daddy of rock falls and we spent a great number of Swiss francs and flying hours searching for the right location. Eventually, I suggested that it should be taken in three separate places: the first a compact crag of relatively easy flying access for the close-ups of the actors, whilst the other two were serious locations where only mountaineers and climbing cameramen could go. Some of the scenes required boulders cascading down relatively easy mountain slopes and the guides with the Second Unit team did this on their own.

The actors' rockfall location was a compact granite pinnacle called the Pel. It required large angle-iron 'shelf brackets' put on its vertical side for the actors to stand on whilst three cubic metres of artificial rock was dropped on them.

Joe, Mo and Ian started drilling the face to take fixtures for the brackets, and above, a protective mesh canopy was made to deflect the 'boulders' from Sean and Lambert. Joe did most of this work hanging free from ropes, like a spider, in a sit harness. It was a strenuous, technical task.

The boys were in their element during the next few days. They were far from the bustle of the film circus. We had a large hopper made up with a bomb release trapdoor fitted. This could be uplifted by the helicopter and opened by remote control and the contents released. Rather than using this directly from the helicopter, we flew it to the top of the pinnacle and placed it on a skeleton-like scaffold frame made by Mo and Ian. The platform was completed to everyone's satisfaction, with its rock-deflecting mesh canopy. All was set for the shoot. We did the first dummy-run on a breezy day and were astounded when the rocks, after falling realistically for about 30 feet, were suddenly catapulted upwards by a strong gust like volcanic bombs, passing the hopper *en route*. This was remedied later by using heavier rocks intermingled with a few real ones and stressing to Fred and Pepino that a calm day was essential for the operation.

I think that we all got a bit carried away during this episode for, with the first fusillade, Fred's helmet was knocked off, even inside his protective cage alongside the ledge. Andrea was perched parrot-like in the wings, in the fall line of the pseudo-rocks. He had volunteered for this position with a bag of 'extra' rocks to throw close to Sean and Lambert's heads, for more 'realism'. Several genuine rocks had been put into the hopper by mistake and Andrea was hit on his helmet by one or several of these, which almost knocked him out. As he hadn't bothered tying on to the face he was very lucky not to have contributed too much realism to the sequence.

As the prop rocks matched the granite rocks of the area perfectly, it was quite difficult to identify the fake ones and the camp crew spent hours scouring the landscape to retrieve them after the shot. I burst out laughing when one of the camera crew jumped on to a large rock which immediately slid from under his feet. He crumpled up in a cursing, undignified heap on the grass. A local primary school group, who had come up the valley on a nature study expedition, joined in the search for these 'rocks', and later we saw them running down the path, brown-limbed, yodelling, to the cable car station clutching huge 'granite boulder' souvenirs.

One of the other locations for the rock fall sequence lay on the main divide between Switzerland and Italy. Andrea had suggested this location: an ugly, intimidating overhang on an east face. By that time, the Ladd Company was increasingly agitated by the escalating costs and it became obvious that we were going to have trouble finishing on schedule. The Mafia were flown up to prepare the rocks on the edge of the overhang. It was a difficult landing on the ridge nearby. A dummy representing the guide had to accompany this live ammunition manoeuvre. I didn't tell the Italian authorities about the project as I knew that we would then immediately be bogged down by bureaucracy. The area below the rock wall is a deep loose gully 1,500 feet deep where no self-respecting climber would ever place a cramponed boot. However, bearing in mind that some climbers lack self-respect, we intended to have a scout round with the helicopter before the rockfall was released.

Arthur Wooster had three cameras set up for the shoot and with Joe, Mo, Martin and Ian at the summit, the huge cascade of rocks was released. The dummy, dressed in duplicate clothes of the guide, was given a one-way ticket with this bombardment. Rocks free-fell hundreds of feet without touching, then exploded at the top of the gully and ricocheted against the gully walls below. There was a pungent smell of brimstone.

The last sequence of this three-part fall was done on the east flank of Pic Morteratsch, the footstool of the famous Biancograt of Piz Bernina. Here, another body was dropped down a steep ice scoop into a gully to an open ice face. The plan was that it would shoot over the top lip of a yawning bergschrund at the bottom of the ice. This time our limestone ammunition was piled up at various points to give the sequence a more sweeping effect, to fit into the end of the edited fall. Rather than risk the dummy snagging on a rock spur, I suggested dropping it from the helicopter so that it would first hit the slope at the start of the steep ice, just after the rocks were released. The co-ordination was tricky. We had three cameras operating. Sitting on the floor of the helicopter with my feet on the skid, secured by a line, I held the 'guide' on a cord so that it dangled 20 feet below my boots. Capito was at the controls. Once I was ready and everyone else primed to go, I said to Capito over the intercom, "If Arthur and the boys are ready, tell them to start rolling both rocks and film."

A few minutes later, I saw the tell-tale puff-puff of rocks gaining momentum high up the face. This was Tut Braithwaite's hoard. Boulders bounced like tennis balls, sending fragments of ice high in the air. Then Joe's heap disintegrated, spreading across the ice in a lethal curtain. I yelled over the intercom, "Dropping the dummy now, Capito. Tell . . . "

Arthur didn't need prompting; the cameras were already running. As soon as Capito saw the falling dummy shoot past his belly-view mirror, he gave the Lama maximum torque, then swung up out of shot. We circled round and, from about 1,000 feet above, we saw the body shoot down the ice for 500 feet and slow up just before it got to the bergschrund. Then, in slow motion, it seemed to slip almost reluctantly over the edge to land on the lower lip of the schrund, 50 feet below. As we circled round, we saw a group of tiny figures far below on the moraine above the Boval Hut. Thank goodness we had sent a warning about the rockfall to the hut custodian on the radio.

Retrieving the dummy wasn't easy. The one which we'd sent down the gully into Italy, in the other stonefall sequence, landed in a frightening place, but it was a relatively easy winching operation to retrieve it. Joe had gone down on the winchwire, and it went like clockwork. However, here, on Pic Morteratsch, though it looked a piece of cake, when I was winched to the bottom lip of the schrund, I found that the body was almost completely buried by debris that had avalanched on top of it. I didn't want to leave it there, for after a thaw it might have been spotted by passing climbers and could cause unnecessary confusion.

With Capito hovering 60 feet above, I became so absorbed in digging out 'Lambert' that it was only when several large rocks thudded into the snow beside me that I realized the stonefall was not yet over — or else another had started. It was obviously not a place to loiter. I managed to signal to Garry whom we had picked up to operate the winch. He peered at me from the helicopter with a look of concern. I felt that friendly come-to-me tug as the winchwire tightened and I was lifted up and out, far above the valley, like an ungainly non-combatant fish. Capito had seen the rocks spilling over the schrund and had taken the chopper up and out even before Garry could warn him. He had seen this because, when helicopter pilots are on hover, they take a 'fix' on some object, and on this occasion Capito had taken a fix on a point just above the top lip of the schrund and had seen the rocks hurtling down.

Before we realized it, the summer had changed its T-shirt for a pullover — there was a chill in the air. Tourists were getting thin on ground and glacier and at the weekends the locals hibernated more than ever. We were now impinging upon the main hunting season. Andrea was growing uneasy; besides being a crack shot, he was an official on one of the local hunting committees. It became particularly embarrassing for him. One morning he was confronted by an angry clutch of *jaegers* alongside the helicopter just when we were about to start the day's work. There were obviously harsh words being spoken in Schweizer-Deutsch.

The pursuit of chamois is a tough sport, involving long stalks in knee-aching terrain. Though I don't sympathize with the motives of the sport, I could see the hunters' point. After a gruelling day to have the *coup de grâce* shattered by one of our passing helicopters could increase blood pressure. The pilots were genuinely worried that we would be shot at by these latter-day William Tells.

The crew was shrinking daily until only technical staff and the Mafia were left with a handful of transport personnel. There were still quite a few sequences to do on the mountains, but there was no way that they could be crammed into the allotted time. The studios were booked back in London. There was only one answer: to return to Pontresina or to go somewhere else later in the winter where we could find snow and suitable mountains. Sean suggested Peru, but we knew there were too many political complications there. Obviously, the problem couldn't be resolved until closer to the time when people were free again. Sean was committed to other projects. Peter Beale reckoned that the footage required could be shot in a couple of days.

Mo departed on a Himalayan expedition and other members of the Mafia started to head home. Betsy and I decided to drive back via Hans Spielmann's hotel in Austria, but we discovered that even this apparently simple exercise had to have the blessing of the film's insurance company, for she was insured for sixteen million dollars. There was still four weeks' shooting to do in the London studios.

We were into the New Year before a convenient week was found for the remaining shots to be done. By this time surveys had been made all over Europe. We considered such far-flung places as Colorado, Ben Nevis and Mexico. In the end, we returned to

View from a helicopter of the Glacier Camp showing the mess tents and bottom right the reinforced 'cornices' on the edge of the crevasse used for the crevasse jump *(photo © The Ladd Company. All rights reserved)*

Sean Connery on face in part of the rock–climbing sequence

Betsy Brantley holds up one of the artificial rocks

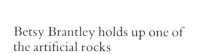

Right: Joe Brown and Tut Braithwaite at work on the crevasse jump below Piz Palu

Below: Eric Jones 'cultivating' the summit of Piz Corvatsch to clean the snow

The simulated rock fall on Piz Pel: Lambert Wilson and Sean Connery protected
by mesh canopy above, with the majority of 'rocks' falling between actors
and camera position

Martin Boysen, one of the doubles, jumping the big crevasse. The cornices, designed by the author, are reinforced with overhanging logs. Betsy Brantley did in fact make this jump *(photo © The Ladd Company. All rights reserved)*

Betsy Brantley with Sean Connery

The author with Fred Zinnemann

Left to right: Norman Dyhrenfurth, Tony Waye, Wendy Leech —
Betsy Brantley's double, the author, Betsy Brantley, Fred
Zinnemann, Giuseppe Rotunno *(photo © The Ladd Company. All
rights reserved)*

Above: Ben Nevis: the summit plateau showing the camp and John Poland's helicopter

Right: Ben Nevis: Charlet skiing in gully

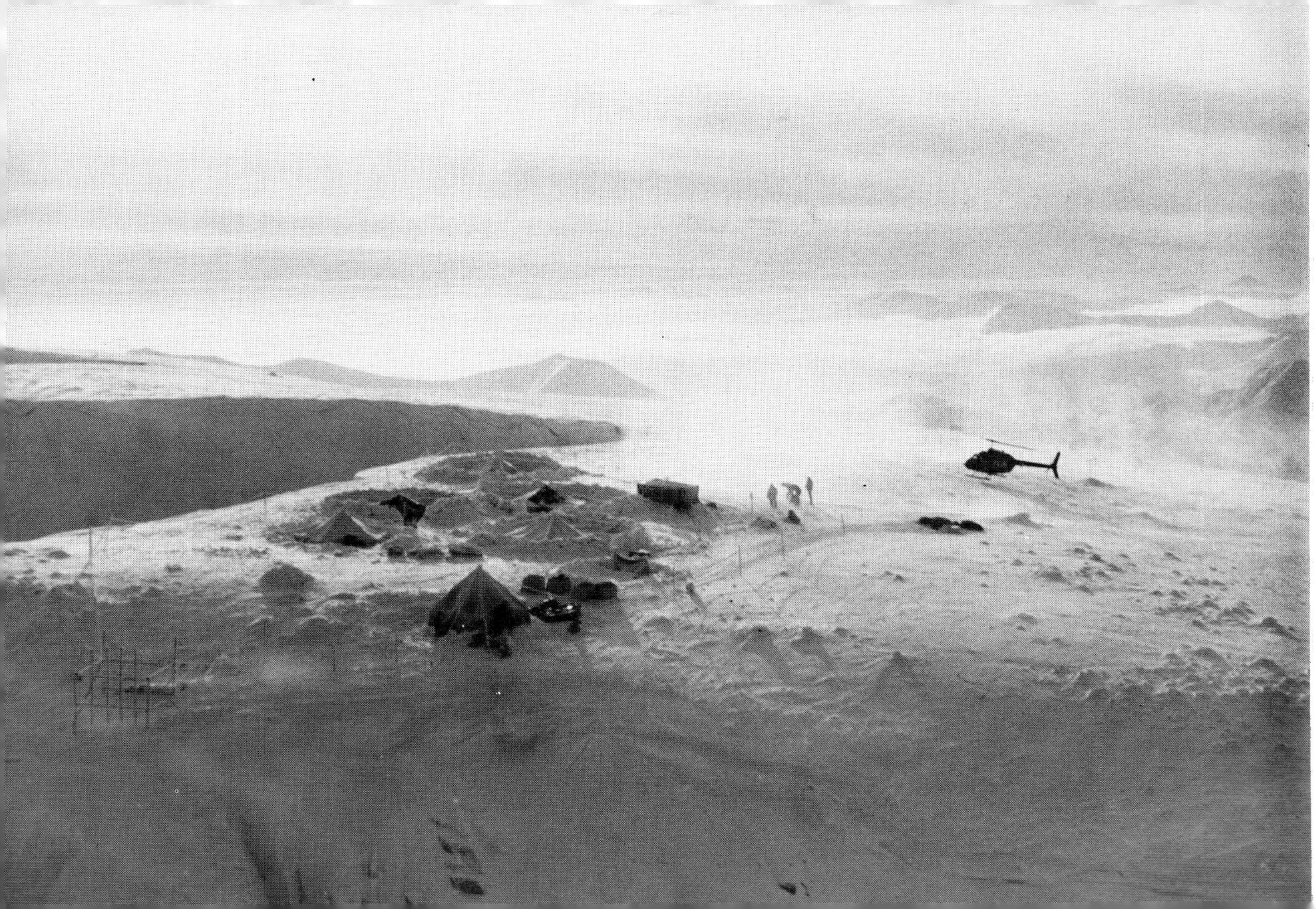

Ben Nevis: tents on the summit

Pontresina. Although the Engadine is very cold in winter, there are also long periods of fine weather.

At the end of January about 20 of us boarded a charter plane at Heathrow, bound for Pontresina. At last we were going to operate with a small close-knit crew and we felt optimistic. There was only a couple of days' shooting to do and Arthur Wooster was the cameraman.

Sean Connery and Lambert Wilson had arrived in Pontresina independently, so we were at work the following morning. The weather was perfect — clear blue skies by day, and, at night, brass-monkey frost. Andrea had already found a couple of the locations for us and we used the Alouette to shuttle people into them. Two of the sites were high and exposed, both to weather and to lump-in-the-throat drops. Sean, lowered into position on one of these, showed consider-able fortitude with only the odd mutter, "Eh, what the hell, Jimmy?" On another occasion, when some of the Mafia were working on the face below, speaking amongst themselves and their conversation, as usual, peppered with invective, Sean observed dryly, "You know, just close your eyes and you'll think you're in a Glasgow shipyard."

Working with this mini-crew was right up my alley and, I felt, everyone else's too. Fred Zee was happy; he was getting his footage. Peter was happy; it wasn't costing the earth. There wasn't the pressure of a large army on the march, or otherwise standing still and ringing up dollars. However, there was the drawback that new ground couldn't be prepared. It's all right for climbers to step out of a chopper on some ice-bound crag in the middle of a Swiss winter, but not for inexperienced flatlanders. The small unit's versatile mobility had this Achilles heel, I discovered.

I had landed with Joe on the summit of Cambrena, looking for a suitable ice slope for a short sequence. It seemed ideal for what Fred wanted. I cut a corridor of parallel grooves in the hard snow-ice to indicate a safe passage between some crevasses. When the others arrived, we told them not to stray off the beaten track, as to rope them all up was impossible. Anyhow, we had previously discovered that to have the technicians strung together developed into a similar situation as when a husky team has an internal squabble, and their traces get hopelessly tangled.

All true artists become absorbed in their work and George Frost, our make-up maestro of boundless talent, was no exception. Here was a man more at home on the pavements of London than on desolate Alpine summits, clutching his make-up bag as if it bore vital dispatches, balaclava pulled down Ku-Klux-Klan-like with his lethal crampons flailing the snow cover. Seeing a flaw in the make-up of one of the doubles, he immediately went to work and, stepping back, outside my parallel lines, to consider his handiwork, he disappeared from view. Only a red balaclava was left with us, George's head still inside it, while the rest of Mr Frost was wedged in a crevasse. Possibly, for once, he appreciated his ample proportions, for it was his girth that did much to save him from a considerable drop into the icy depths.

All this time Fred Zee was ecstatic. The mountains seemed to act on him like an elixir. Normally dynamic and alert on the hills, he now seemed charged with youth. Indeed, I felt that here was a realm that he had been deprived of for years and now he grasped it and its magical properties like a connoisseur of fine art absorbing an Old Master. He always wanted to be with the first up in the morning and with the last trip down. As some of these locations were in remote and often difficult places, hard enough for even an experienced mountaineer to get back from before dusk, he caused us a certain amount of consternation. If the helicopter had broken down at that late hour, we could have been faced with a serious situation.

The most memorable part of the final shoot in Switzerland was not concerned directly with the film. Joe, Mo and Martin had, prior to arrival in Pontresina, fixed ropes on the end part of the East Pillar of Piz Palu. Owing to lack of direct winter sunlight on the Pillar, we had been unable to use this location. As a matter of fact, I was suspicious of its suitability when I first arrived and I had asked Mo to go up the valley from Pontresina towards the Diavolezza cable car station and sit in the sun amidst the pines with a couple of cans of beer, and observe shadow and sunlight on the face. He later confessed that this was the best job on the whole trip, but judging from his state on return, it is doubtful if he could even have seen the peak at all, and he had certainly taken more than the prescribed two cans of beer.

A couple of days before, Fred had asked me if it would be possible for him to do a climb before returning to London and I had suggested Piz Palu as the fixed ropes had to be taken down in any case. Martin was standing beside me as I talked to Fred about this, on that final day of the shoot.

"Would you like to come up Piz Palu with us, Martin?" I asked. "Fred is keen to get a climb in."

"Sure, that would be great. Still some life in you yet, Fred!" Martin laughed and poked one of the world's most eminent directors in the ribs.

"But are you sure I wouldn't hold you up?" Fred felt a tinge of doubt.

"Not at all," I returned. "Just what we need. We can take the helicopter up to the Silver Saddle and climb from there. That will save that Russian Roulette walk across the glacier, for the crevasses are like unemployed graves now they're hidden in winter snow."

"You know, I feel that climbing with Martin and you is like playing tennis with Borg and Nastase."

"Hey, I overheard all that." It was Peter Beale. "I'd like to go as well."

"I'll take you," Joe volunteered. "It would be better if both Hamish and Martin look after Fred."

Next morning, those who weren't mopping their brows and lubricating their palates after this final round with the mountains were to go skiing or, in our case, climbing, for our charter plane wasn't due to arrive until the following day. It was the usual, rather monotonous picture-perfect morning, standard immaculate blue and white. Joe, Martin and I piled in beside Ueli and we took off; the whole floor of the Upper Engadine fell beneath our feet as we rose in the helicopter. In front of us, Piz Palu, with its three pristine buttresses, looked like a creation of a master pastry-cook. With us, for the ride, came the local police mountain-rescue guide, who had worked with us on our last visit. We put down on the Silver Saddle and the Alouette returned for the next lift. With the helicopter parked on the saddle, Martin and I set off with Fred roped between us. Joe and Peter followed a short distance behind. Ueli was itching to come along, but as they were on rescue stand-by, he couldn't leave the helicopter.

The ice was superb and the conditions far better than in summer. We were making fast progress, crampons biting into the ice with that pleasing 'clomp' which gives one the superb feeling of security and purpose. Fred, who was going like a teenager, not a septuagenarian, kept apologizing for his slowness.

"For goodness sake, Fred," I said, exasperated, "neither Martin nor I want to go any faster. This is fun!"

At this point, we had slanted on to the edge of the ice ridge. I was up ahead with Fred on a short rope, whilst Martin was in the slips just below Fred's crampons, making sure that our director didn't make any false moves. From time to time when I had moved up a more awkward bit and was belaying Fred, I could see the helicopter perched below on the saddle, like an exotic red bird. Just behind it was the face where earlier two bodies mistaken for our props had been found.

Soon there was nowhere else to climb; we had reached the summit of the mountain, and stood well back from the 50-foot cornice.

"You know, Hamish," Fred said, "this is one of the great days of my life."

"Well, you did well, Fred. Not many people do a winter ascent of Piz Palu, and even fewer at your age."

The following morning we all congregated at the well-remembered airstrip.

It's my view that those who fly regularly in small aeroplanes are lucky to be alive, especially in the fickle weather which we have in Northern Britain. Obviously, the pilot of of our charter plane was concerned. It was not the same man who took us out from London. I personally thought that the previous pilot would still be on tranquillizers. This pilot came into the waiting room of the airport and glanced nervously at the luggage. He counted us for the second time, then went furtively off to consult his various tables and I felt my heart sink. A short time later our baggage was collected and we made our way across to the plane. The strip was clear of snow and, all around, the mountains seemed etched out of the blue.

Ueli, Capito and Andrea said *au revoir*. We'd had a lot of fun and a few frightening moments together over the last four months. The pilot came over and entered the plane.

"I'm sorry, gentlemen, but can you all return to the waiting room? Everything has to be reweighed, including passengers." I had already heard, when I arrived at the airstrip that morning, from one of the helicopter mechanics, himself a pilot, that when this particular charter plane had landed earlier, it had done so in an alarming way, bouncing high in the air after touchdown. That had not engendered any confidence in our latest pilot.

After being weighed, we returned to the plane and were standing round it, waiting to go on board. I observed several puddles of oil beneath the engine, and the pilot's answer to my nervous enquiry was hardly reassuring.

"It's a bit past its prime, sir. We all get old." However, Joe Brown, whose power of observation rivals the hoody crow, bent down beneath the fuselage and picked up a fragment of aluminium alloy. He inspected it with a puzzled expression, as if it was a dubious belay.

"Hey, look at this, fellas!" By then he was down on his hands and knees to examine one of the wheels. "Hey!" he shouted. "The bloody wheel's broken up."

It had. Obviously a casualty from the morning's impact. A buzz of excitement, and then a spare wheel was found from somewhere in the innards of the aircraft and hurriedly substituted for the faulty one. The concern of the pilot proved to be a reflection of his caution, for we took off with only a billiard table's length of tarmac to spare.

Someone mentioned casually, as we were flying high over the spiky Alps, that only one of the wheels of the undercarriage had retracted. At least, I consoled myself as I accepted a large, and, I felt, medicinal brandy from the regulation blonde hostess, there was a longer airstrip at Heathrow . . .

O.B. OR NOT TO BE

ONE OF MOUNTAINEER Don Whillans' favourite saws is: "There's no fool like an old fool." This is certainly the case as far as my association with Ben Nevis is concerned.

Back in 1962, I made an ice-climbing film on the mountain for the Air Ministry, with the RAF Mountain Rescue. This should have taken five days and, instead, took four months. In the winter of 1978, I spent three months on the mountain attempting to make a documentary film with *National Geographic Magazine*, starring two friends, Yvon Chouinard and John Cunningham. John Poland was our aerial taxi driver in his very first Jet Ranger. Not only was the film never completed satisfactorily, but the helicopter had a three-day forced bivouac perched on the flanks of the mountain with ice in its compressor. A frosted mechanic made his début on Scottish mountains to take it to bits, then reassembled it in a force-ten gale. In the meantime, John's poor machine was tethered by six climbing ropes anchored to five-foot-long steel spikes. Even with these restraints, the helicopter was still attempting to take off without assistance from its engine. On the very last day of this circus, an avalanche thundered down Number 3 Gully, on top of Yvon, John and myself. John and I managed to get clear, in a sprint, wearing crampons instead of track shoes, but Yvon was snatched up in the first snow wave and bundled down 600 feet. He survived, but a great deal of costly camera equipment was committed to the snows.

Some years later, producer Mike Begg and I did catch this hulking pile of rock and snow unawares when we managed to complete a short snow-climbing film for BBC2. After that, my luck reverted to normal — with the balloon flight.

When I first mentioned to Mike the possibility of attempting a live O.B. in winter from the summit of Ben Nevis, I could have bitten my tongue off, for I should have known better. Later I discussed the idea with Mike's boss, Phil Lewis. Phil is a level-headed enthusiast, tall, bearded and pipe-smoking in the Alan Chivers mould. Once he has the bit of a programme between his teeth, he doesn't let go. We were in his office at BBC Kensington House.

"How can we get a signal off the summit or from the climb, Hamish?"

"Just one relay point, Phil. From the top of the mountain near the old Observatory site — where Tower Ridge finishes — down to the Allt a

Mhuilinn Glen, then on to the south-west end of the Great Glen, near Fort William."

Although the advent of satellites now allows a live O.B. to be mounted almost anywhere, the ghastly weather on Ben Nevis makes a broadcast from there more difficult than a major Alpine or even a Himalayan programme. One can't alter climatic conditions. On Ben Nevis they are bad. The policy of the BBC with live climbing broadcasts is to have successively more interesting and ambitious projects. Ben Nevis offers plenty of interest, but weatherwise, no self-respecting bookmaker would ever give us an outside chance.

The Ben Nevis idea was, of course, conjecture on my part, but I had worked with BBC engineers for a number of years and felt that they had the know-how.

Phil liked the idea of the Ben.

"Let's have a look at it," he urged in his positive way. "It sounds interesting."

Several recces later, it was established that the proposition looked feasible and plans were laid for the following February 1982. As I reasoned that we might as well be hung for a sheep as a lamb, I suggested that we get an extreme skier to descend an ice route on the north face of the mountain.

I had been told of the exploits of Jean-Frank Charlet, the mountain guide and extreme skier from Chamonix, who came from a long line of guides. Jean, himself a brilliant climber, had already done some of the great classic snow and ice routes in the Alps and in Scotland.

I contacted him, suggesting 'Good Friday Climb', a Grade III winter route, and sent him photographs of the face. He thought it feasible, but of course he would have to climb it before committing himself to the boards.

The cost of an undertaking like this would have been beyond the pocket of the BBC, so Mike asked the Royal Marine Commandos if they would like to participate 'as an exercise'. The Marines normally do winter survival expeditions on Ben Nevis and in Norway, anyway. They agreed. They also had Arctic tents in their stores, which would be vital for technicians and equipment on the summit.

The climbs chosen were two of the hardest on the mountain, Psychedelic Wall and Gardyloo Buttress, both Grade V climbs. Murray Hamilton, who was on the 'Freak Out' programme, was to team up with Captain David Nicholls, a Marine officer, whilst Joe Brown was to climb with Betsy Brantley. It was Joe who had taught

Betsy rock climbing in Switzerland for the Zinnemann film the previous year and she had been ice climbing with me that winter. Though their route, Gardyloo Buttress, is one of the top-grade climbs, we thought Betsy could second it, for she had proved to be a talented climber on ice as well as on rock. Both Mike and I also felt it would be good to have someone on the box who wasn't an ace mountaineer, as it could show people that such exploits are possible for 'normal beings'. John Poland's helicopter firm, P.L.M., was to do the milk run up the mountain with a back-up for heavy lifts by a naval Sea King helicopter.

The plan was to have several camps, other than the summit mini-village; one at the link point, immediately to the north-east side of the mountain in the Allt a Mhuilinn, and three further up the valley in the shadow of the Ben's cliffs. These higher camps were for radio communications and camera positions mainly to cover Jean-Frank Charlet's ski spectacular.

The gods were against us from the start. We had probably chosen the worst winter for donkeys' years to attempt this broadcast. Rain had started the previous July and was followed by the worst gales in living memory, which had continued unabated over Christmas and New Year into February. It would be obvious to anyone misguided enough to visit the Highlands during this period that there was a genuine necessity for the national footwear — 'wellies'. Very little climbing had been done and, as we prepared the equipment to be flown up the mountain, we were not without misgivings.

Jim Maiden, who was the engineer in charge of the 'Freak Out' programme, was also the man behind the equipment on this present project. Roy Carpenter, Deputy Head of BBC Outside Broadcasts, was in overall charge of the technical planning of the operation. I suppose what this meant in BBC pecking order was that Roy could lose his head as well as his legs if the programme was a disaster, for after one of our surveys up the damp and peaty Allt a Mhuilinn, Roy almost disappeared in a quagmire resembling a soggy Black Forest gâteau, and thereafter became known as 'Roy the Bog'.

Mike, Director of the O.B., was also going to produce a documentary film of the operation. As he wryly observed: "I may as well have a record of my last days with the BBC!" For my sins in promoting his *hara-kiri* dream I was made Producer; I too was tied to the mast.

As well as the Glencoe Rescue Team acting as Sherpas, we had a handful of other well-known climbers there. Rab Carrington and John Yates, two leading Himalayan climbers, were to work with Mike's documentary film crew. Both had been on the Zinnemann film in Switzerland the previous year.

There were two commentators: Eric Robson and Ian MacNaught-Davies. Ian is a mountaineer who took part in several of the early O.B.s as a climber. This time, he was going to talk from the top. But as he was then suffering from a bad knee, he felt he would require the joint effort of our Rescue Team to carry him up if the helicopter couldn't make it, thereby establishing a record as being the first person to be carried up to the summit on a stretcher!

Sergeant Dave Lazenby was in charge of the Marines on the summit. Contrary to his name, he was a hard worker. He could have been one of those warm-hearted characters out of 'Barrack Room Ballads', generating good humour and bonhomie. The first fine day, which was really a blinder, their camp gear was flown to the top. The summit was covered in a heavy coating of fresh snow and they managed to dig their circular Arctic tents deep into the surface. Viewing them from the helicopter reminded me of dart boards on a pure white wall.

John Poland and David Clem did some aerobatic helicopter flying in the gales that tore at the mountain over the next few days, snatching every opportunity to fly through holes in the cloud. Each grey wet dawn saw more equipment and trucks squelching into the base site, a caravan park just outside Fort William. With eight days to transmission, only a fraction of the gear had been taken up the mountain. Some of the Glencoe team were working with Marines in the Allt a Mhuilinn, establishing camps for cameras and communications. Already a Marine contingent were living at Link Camp under Corporal Wilkie, a lighthearted extrovert of boundless energy. The Marines were amazed at the number of climbers who trekked past their camps. The Scottish mountains, especially Ben Nevis, have an international reputation for ice climbing in Arctic conditions, drawing mountaineers from all over Europe, America, and even Africa for the winter sport. I regret to say that often the end result of their travels is merely an opportunity to squelch through bogs in thick mist or be hammered by gales and blizzards. There is a high accident rate in this area, especially during heavy snowfall, as it was just then. That day, the 15th of February, David Nicholls was doing a training climb on the Ben, though conditions were miserable. He described what he saw when he came down towards Link Camp:

"I looked up right and saw a great powder avalanche coming out of

Castle Gully. Initially, I thought it to be a vast powder cloud. Then, as the cloud descended, I saw that there were lots of climbers in the gully and they were caught in this avalanche. Then I saw three of them swept down the gully into the first snow bay. Then, to my horror, they fell 150 feet straight over a vertical rock face. We went up and, digging into the snow, found two climbers. Sadly both were dead."

We had a hard day on the hill, with flying restricted to only one flight. By the time I returned home, I found I'd missed a rescue call-out in Glencoe. Actually, it was in the Mamores, a range of hills between Fort William and Glencoe. Some of our team located the fallen climber, but as he was dead, they left it until daylight to bring the body down. On my way to Ben Nevis the following morning, I stopped by the team rendezvous point. Things were running pretty smoothly, and they brought the corpse down later without incident.

During another 'porthole' in the atrocious weather, a Naval Sea King helicopter arrived from Prestwick and, together with the two P.L.M. Jet Rangers, shuttled mounds of technical paraphernalia and supplies for the Marines on the summit. When the weather cleared up there the Ben smiled, as if it had always been a model child. It seemed unbelievable that, in ten minutes, a blizzard could be swirling in tantrums from any point of the compass. One did that night, and several feet of snow fell, threatening to smother the camp. We were still in radio contact with Dave Lazenby; he and his band were cheerful enough. Dave was only worried that they wouldn't get their delivery of cigarettes and daily papers.

By first light, all his men were out with shovels, digging the tents clear, some of which had collapsed with the sheer weight of snow. Also our fixed ropes, which led from the plateau down to the climbs, were now covered with about ten feet of powder which had funnelled down the gullies; it was unlikely that we could dig them out. Things were not all beer and skittles at Link Camp either, and several of the tents had shredded in the storm. Sometimes it was only possible for the helicopters to get part-way up the mountain on summit trips, and from there we had to go on foot.

Any technicians going up had to be escorted by climbers, as it was essential to travel on a compass bearing across the wide exposed back of the mountain, where visibility was sometimes down to three feet. To one side are the great cliffs and to the other the steepening slopes dropping to Glen Nevis, reamed with ugly gullies eager to collect stray climbers. One of these, Five Fingers Gully, can funnel lost climbers down any one of its digits on to the steep uncompromising ice pitches lower down.

Back at base there was much shaking of heads and eating of bacon-butties. The caravan site looked like an earth-worm convention: a mess of cables interspersed with trucks and pools of water. It was raining, with a wind that, even down here, was blowing at 50 knots. Jim Maiden, the engineer, was getting decidedly agitated, and there were mutterings of pulling out. Both Mike and I felt we were in for a penny, in for a pound, and were not going to be short-changed by a Ben Nevis suffering from wind. Months previously at a production meeting, Roy Carpenter had asked who would decide when, in the event of catastrophic weather, the O.B. would be aborted. This decision, from the safety aspect, fell on me. But Mike and I had no intention of backing out now. As soon as the crew were exposed to danger, or rather, before then, was the time to pull out the plug. All we needed were two clear days to get the rest of the gear up and we would, as they say in space jargon, be in a 'go state' to transmit a picture from the summit.

Flying was no longer fun (to me it never is, especially on the Ben), and John and David would toss a coin to see who bought the first flight of the day, always the worst.

One morning, David and John were standing alongside their machines, and John took out a 50-pence piece.

"Toss for it, Dave?"

"Okay, heads I go, tails you go."

This always created laughter among the climbers and BBC crew. It had, I suspect, the sort of macabre humour one would associate with a wartime air offensive.

The coin spun and dropped in the grass.

"Oh — me," David said with resignation.

"Actually, I think I've lost the toss, Dave," John said with a hollow laugh.

"Why?" Dave asked, puzzled.

"Well," John returned, "I've got to fly as well. Hamish wants a recce — the back corrie; he doesn't want to come himself, though."

More laughter.

Mike Begg took a recording of Dave's flight up the Allt a Mhuilinn, which was by no means an exceptionally risky one, yet it depicts what a pilot has to contend with in the Scottish mountains in winter.

"What's your advice to a young pilot who wants to fly here in winter, Dave?" Mike asked.

"Ha ha — tell him to go and play cricket in the Bahamas — I don't know. It just amazes me how long we can survive. If you think of the number of heart-stopping instances you get in winter; it's proportionately much higher than in summer, mainly because of the climatic conditions: gales, rain, snow, icing and bad visibility. Also, the turbulence. Well, you can feel it as we come up over this ridge. Oops, there's one there, that great lurch. And it's this horrible sensation as you go sinking towards the ground with the lever up under your armpits. And you're just waiting for 'will we stop or will we crash into the thing?' I still haven't managed to get over that even after seventeen years.

"Now today, we've got at the most twenty knots of wind from the south-west; you can feel that we're being thrown around like a cork in a bucket. There's no rhyme or reason to what happens in this glen, it makes its own weather. You can even get flat, calm days in winter and there's wind swirling round and round here like nothing on earth. Well, feel this lot . . . It makes life interesting, I suppose."

At this point, the helicopter dropped about 50 feet.

The storm didn't take any time off and it was only with luck that the helicopter managed a mercy trip. By the time they got the machines wound up when they thought a clearing seemed imminent, it would sock in again like a wet dishcloth. Dave Clem managed to sneak in one trip, and as he said: "We were lucky. We got a break and came in. It cleared off. There was about two feet of fresh snow; the summit camp looked like a buried Everest expedition, but we went in and landed. Snow spilled round, then the cloud rolled in and the whole bubble of the helicopter and the blades iced up, just like that. I couldn't see what the hell was going on, I didn't know whether to stay there, shut down, and clean everything off, or what. But then it cleared slightly, so I just set a hundred per cent torque, watched the instruments and climbed up on top and came down . . . I don't know."

This storm, which Dave managed to cheat with a delivery of goodies to the Marines, was a mile high with winds of over 90 m.p.h. It lasted 72 hours and caused devastation at all the camps.

I had just got back home on the second day of the storm, had a bath and was looking forward to getting my feet up by the fire, when the phone rang. It was another call-out; Willie Elliot, our local National Trust Ranger and gamekeeper, gave me the news.

A solitary climber was overdue. We knew where he had set off to climb — Stron na Laraig, which was a long way from the road — and he would have had various descent routes to choose from, which I knew would complicate things. It was now a case of alerting the team and requesting a helicopter, a Sea King from RAF Lossiemouth. But we knew at night it was like looking for a snowflake on a glacier if he couldn't signal to us.

I went up with the helicopter when it arrived, but the crew were psyched-out by their flight in total darkness between 3,000-foot-high mountains (it was an ink-black night). So was I after 20 minutes up in that deafening roar and absolute blackness. Eventually they gave up and landed me back at my home.

We then went in on foot, but did not find the missing man, so activities were abandoned until first light.

I was up in the grey slushy dawn with a large party, which was now swelled by an RAF team and a Lochaber team. Later that morning the climber's body was found by one of the rescue dogs and airlifted to base by the RAF chopper.

Dave Lazenby's summit camp was a Shangri-La to overdue and benighted mountaineers who reached the summit with about as much idea of where they were as a Glaswegian on Hogmanay. Usually, after a cup of hot chocolate, they were sent on their way, for there was little room in the Arctic tents. By now several of these had been wrecked by the storms and the rest were crammed with electronic exotics. In the fangs of one particular blizzard at 11.30 p.m. there was a noise at one tent's 'front door' (not exactly a knock, though it was frozen hard enough for one). Anyhow, the presence of three climbers was made known — they were lost. After the regulation Samaritan brew, Dave suggested that they should bivvy in one of the tents for the night, as the storm seemed to threaten even the mountain itself. However, the climbers were afraid that a search party would be called out. Dave told them that the descent route to the Allt a Mhuilinn via Number 4 Gully was the third gully on the right. Several hours later the climbers returned, encased in ice, as they had failed to find the descent route, and could they now take up his B & B offer?

Engineers who had never been on the mountain before, gamely donned their protective clothing, crampons and ice axes, and followed in the footsteps of the Glencoe Mafia. They worked in atrocious conditions: blue fingers and dripping noses, assembling intricate units in competition with spindrift in an atmosphere comparable to that of a flour mill which has just been bombed.

I put Ronnie Rodger in charge of the Allt a Mhuilinn Mafia squad, and being dedicated, he felt that he had to show the flag and take the first flight up this sinister valley each day with either John or Dave. One particularly blustery day, when even at the car park the wind was rocking the trucks, Dave and Ronnie climbed aboard the Jet Ranger to check the camps. The Marines occupied only the Link Camp in the Allt a Mhuilinn. Dave took off like a learner buzzard on its inaugural flight. I got involved with some other work and only later heard an account of the trip from Ronnie.

"After the initial turbulence, it wasn't too bad as we passed into the valley and approached Link Camp. The commandos came out to greet us. Dave circled round and continued up the valley. Over the next camp, one of communication tents, he took the chopper in a wide, no-nonsense turn, which seemed to go disastrously wrong. Suddenly we were pointing to the ground, tail up, bubble down, as if we had been smashed by a giant table-tennis paddle. This was just crazy flying — I thought, what the blazes is he up to? Just as we seemed to be destined to mess up the large snow-covered boulders, he miraculously pulled out of the dive, and giving the machine maximum power, got clear." It had been a tremendous downdraught.

That was the last flight of the day. But two days later, Ronnie was moving this same camp across the stream, or rather where the stream flows in warmer times. With him was Peter Weir and some of the Marines. As John came in with the wire rope dangling from the stomach of the helicopter, Peter clipped on the load for this mountain flitting. However, someone had unwittingly left some sheets of polystyrene lying around without any restraining boulders; these took off in the downwash of the helicopter. John, not knowing what to make of these white airborne apparitions, released the load in the emergency and swung away. In so doing, the heavy mace-like hook hit Peter on the head, making a nasty gash. He was flown down to hospital to have it stitched.

Now every tent in the valley had been blown down except the one the Marines were living in. Some were beyond repair.

Jean-Frank Charlet arrived with his wife and baby, and the next day we arranged to take him up Good Friday Climb. Jean-Frank exuded vitality and enthusiasm and was overjoyed at meeting so many of Britain's leading climbers.

Unfortunately our performance would have deserved a mention in the *Book of Heroic Failures*. The previous night, Joe and Mo had been to a party an hour's drive away, which had expired at dawn. They had left early, at 7 a.m., but still looked as if they had been inspecting sewers for the past decade. I myself had befriended one of those potent winter bugs which invade the Highlands each year, and this manifested itself in violent diarrhoea, with no early-warning signal. Valiantly, we set off with Betsy, dragging on the fleet heels of this super guide.

He soloed the route, followed by the sick and infirm.

"Yes, Hamish, I can ski it all right," he told me at the top, "only with two short abseils, but these can be very quick, with my skis on."

"If the snow is still doubtful when you do it for real, Jean, you can borrow an avalanche bleeper."

"Good idea. But tomorrow I think I'll have a — how do you call it? — trial run."

The following day was fine and the helicopters buzzed between the car park and the summit like bees who had found a lake of nectar. All work stopped when Jean stood on his skis poised on the summit cornice. Will Thompson, a member of the Glencoe Mafia and himself a keen skier, had fixed a couple of short ropes down the two ice pitches. With a whoop, the mad guide was off. He had told me that he could ski a 50-degree slope in a gully three metres wide. Any mistake in his timing would have meant death over a rock buttress which plunges hundreds of feet to one side of the descent. Mike and Ian Kennedy, his whiz-kid cameraman, were up with John in the helicopter recording the run for Mike's documentary. Jean wore a radio mike and we could hear his commentary. To begin with, he was obviously keyed up, then after the second abseil it was all whoop and swish, lightning-fast jump turns like a conductor's baton thrashing the orchestra in an *allegro*; then down into the wide basin below Gardyloo Gully, where appropriately enough the French term *Garde de lieu* was given to the gully adjacent to the old observatory, down which all the rubbish (in the days before environmental pollution was invented) was heaved. It was a term used in Edinburgh in the fifteenth and sixteenth centuries accompanying discarded slops chucked out of windows each night.

The toll of dead climbers still mounted in the storms, and its resultant aftermath of dangerous hair-trigger snow slopes. The gullies were particularly hazardous with many soft slab avalanches. In these the snow breaks away in a great patchwork quilt, which in turn often erupts into an explosion of powder. Or, if it is hard slab, it breaks up like a giant jigsaw puzzle with individual chunks varying in size from crazy-paving-stone dimensions to that of garage doors.

On the Thursday before transmission, Dave Clem was flying Ian Kennedy's documentary crew up the mountain when he spotted something. Dave is an exceptional pilot. He also has astounding reactions and, like a peregrine, sees everything from aloft.

He said afterwards: "I thought I spied some orange-coloured equipment lying in the snow, but on closer inspection I realized that it was two bodies."

Dave was in a quandary; he wanted to find out if they were still alive, but couldn't land just there. Instead he eased the machine alongside one of the bodies. It was definitely dead. He then moved sideways, and touched the other fallen climber gently with the skid, peering at him through the Perspex floor of the bubble as he hovered. He too was dead. *En route* to base, he gave the news over the radio.

Donald Watt of the Lochaber Rescue Team went back up with David and, landing a short way off, went over and recovered the two bodies.

The depressions which had queued up in the Atlantic showed remarkable solidarity, and with the determination of that dedicated union, refused to relent. Whenever it was flyable, engineers would be taken up as far as possible, then it was 'walkies' in Indian file; a knee-deep trudge to the summit for their deep-freeze shift. Already over two tons of equipment had been relayed up the mountain, but there was still a further ton to go. Generators were running night and day at the top camp to help prevent vital equipment dying from exposure, but the generators were now throbbing under the surface in their own little snowholes and, like everything else, had to be dug out several times every 24 hours. At least they weren't causing noise pollution.

Everyone was worried now, with only two days to transmission. The mound of equipment absolutely necessary for the top camp hadn't diminished. Dave Lazenby's summit commandos had a Nansen sledge and this was being put to use for hauling up hardware which had been taken as far as the bad-weather helicopter ceiling. Some gear was even flown right round to the other side of the mountain as there was a temporary clearing there, but it was a dangerous landing on the narrow Carn Mor Dearg Arete, a place subject to violent winds, so only one trip was made. More of the

Glencoe Rescue Team as well as other climbers were recruited to man-carry the boxes of tricks.

Mac the Tele did in fact claw his way to the top under his own steam and on one occasion I heard him chant as he did a King Wenceslas through the drifts . . . "Tomorrow, tomorrow, I love you tomorrow, You're only a day away."

Though the top of the Ben is probably the nastiest summit to fly round in the British Isles, the Allt a Mhuilinn certainly seems to have the edge on it for turbulence. That ill-fated pair, Dave Clem and Ronnie Rodger, were doing one of their good shepherd flights to the camps in the glen. Dave describes the daily morning flight with his usual understatement:

"We landed on the pad in the upper valley. Whilst we were unloading I saw a wall of spindrift go over Camera Five camp, obliterating it. I watched it go across North-East Buttress, then turn straight for our helipad (actually a ledge a few feet square). It hit the side of the helicopter, must have been a gust in excess of seventy to eighty knots, and physically turned the helicopter on the ground, graunching the snow shoes on the rocks. Ronnie fell over on the snow-covered scree, hit by the fuselage. It was at this point that we decided to give up flying operations into the Glen for that day."

Indeed I sometimes thought that by now the objective of the exercise was direct combat with the weather. Time was running out. Helicopters couldn't fly except to stagger up the first two or three thousand feet of the mountain. As Betsy said in one of the interviews, "The weather seems to be the leading lady in this project."

The BBC crew were all volunteers, yet groping their way up to the top was more than could be reasonably expected of non-mountaineers. Even for the climbers, it was a trial by fury. After one such slogging place-your-boot-in-that-hole session, Joe Brown described the conditions to Mike back at base.

"It's white-out conditions most of the time, so you can't see, and can't tell whether you can see even a yard at times. Situations like this are very dangerous, because you're never sure if you are going to step over a cornice. It's also blowing like hell, which is very painful, and carrying loads in such conditions is desperate. You're being continuously blown off balance, being hit in the face with spindrift, which really hurts . . . But apart from that, it's pretty good."

The wind had been gusting on the summit at 100 m.p.h. This, coupled with low temperatures, made a very potent killer. We just couldn't take risks with the BBC crew. They had to be escorted everywhere and round the summit camp we had two lines of rope fixed to steel stakes to prevent anyone dropping off. On two sides were crags of over 1,000 feet to the Allt a Mhuilinn. Large meringue-like cornices — all puff and no substance — made these edges even more dicey. Engineers had to resort to boiling some of the equipment in water in order to thaw it out, and we didn't have to worry about burying the hundreds of cables that snaked round the camp: Mother Nature had done that for us. Now, some of the circular Arctic tents had only their tops protruding from the snow, and they were nine feet high. It was like a latter-day Pompeii, covered in a white lava. One had to descend to gain entry to any tent, even the loo was buried in the depths; a cold silent place where spindrift-laden draughts sought out exposed crannies, tender and unprepared for such treatment.

Next day, I was with a Sherpa gang, comprising Will Thompson, Betsy, Ronnie Rodger and myself. The weather had been turning up the volume as we ascended. I knew that we couldn't be very far from Camp Lazenby; we were steering by compass. Visibility was about four feet, or perhaps I should say the distance of the two feet directly in front of you. The wind was so pushy that it was difficult to stay upright. There were misgivings about the advisability of going on. Ronnie voiced this doubt: "Do you think we'll make it?"

Will replied, or rather shouted, into the gale, "It's no' far now, let's bash on."

"Okay" I yelled back. "But let's put our crampons on, otherwise we'll whisk off like ice yachts over Number 3 Buttress."

"Good idea," someone called, but I don't know who it was, for the words were grabbed by a gust which just about took us as well.

As we knelt on the ice and put on those tacky lumps of metal with fingers that didn't respond to signals, Ronnie straightened up and I heard his startled cry:

"Good God, Hamish, the two attaché cases have gone!"

Ronnie had been carrying these connected by a strip fluorescent wind marker which we use for attracting the attention of chopper pilots and to indicate wind direction. He had used it as a harness so that he could carry the cases with the sling.

It was a situation of 'now you see them, now you don't'. We were on an ice slope which angled at about 30 degrees towards where we thought Number 3 Gully was. Of course, we couldn't see a thing.

Ronnie went down like Orpheus into the grey nothingness and a short time later loomed back.

"No sign."

"Let's go on," Will said.

"Right," I agreed, casting my vote, "or we'll stay here for keeps."

We made it to the top and back again. At the caravan site Ronnie told Jim Maiden about the loss of the cases.

"Can you describe them, Ronnie?" Jim asked.

"I can," Ronnie replied with his usual exuberance, wiping the melted snow from his chin, "One was marked 'Selwyn's Pandora Box'. That must be Selwyn Cox's, the cameraman who does some climbing. It had his cans inside, and some odds and ends. The other case had those board things in it — you know, kind of action boards or something."

Ronnie recalls, "Jim went ashen-faced", and I thought, 'The tight fisted old bugger, worried about a few old boards! And Selwyn's cans could easily be replaced.'

Next morning, Ronnie phoned me up early to get the day's work programme.

"Hello, Hamish. What time do we start today?"

"About eight o'clock. It's back to the Captain Scott routine, only we don't even have any ponies. Incidentally, Ronnie," I tried to sound casual, "do you know the value of the attaché case with the 'cheap boards' in it?"

"No. Tell me, Hamish." Ronnie's voice had a trace of concern in it.

"About twenty thousand pounds," I said. "See you up at the caravan site, fella."

Ronnie made an all-out effort to locate the cases, worry is often an excellent motivator, and indeed he found them. There was just a corner of one case protruding from the snow; the other was completely buried in a drift. They had stopped short of going down Number 3 Gully, otherwise they wouldn't have been exhumed until the late spring. Apparently the boards are used for lining up the big TV cameras and were unique, being the only ones in existence for these particular cameras.

When Ronnie was married a couple of months later, Mike Begg gave Ronnie and his wife Karen an obsolete camera board as a wedding present, mounted on a plaque with a plate on it stating:

BEN NEVIS 1982

To Ronnie Rodger

Who rescued £20,000 worth of BBC Camera gear by making a hazardous descent of the North Face in ferocious conditions at considerable risk to life and limb.
Serves the silly bugger right.

Mike's humour didn't stop there, he had a T-shirt with 'Blame MacInnes' printed across it in bold letters.

Next day, again in an off-white hell, the convoy of figures trampled the man-made ditch to the summit. It was a situation akin to prospectors crossing the Chinook Pass, only in worse weather, and there wasn't a sign of anyone carrying a piano, though every other shape of load seemed to be in vogue.

On the summit the BBC engineers worked like low-temperature ants, and on the Friday before transmission succeeded in transmitting a picture from the highest point in Britain. Everybody was ecstatic about this; at least we had achieved something.

Tomorrow was to be the big day; the question was, would we make it?

Everything seemed buried in this suffocating snow, even our two chosen climbs. Fortunately, they were so steep that much of the snow cascaded off them. To climb was possible, but to televise it was another matter, visibility was often down to ten feet, even less on occasion. We desperately neded 24 hours of good weather. Everybody was up early, and again it was one foot in front of the other. As we plodded up over constantly filling furrows, buffeted by a south-westerly gale, my thoughts took me to that simple barometer, a man and woman in a mini-house with two entrances. I tried to remember who it was that came out of their miniature home during depressions, the man or the woman. I recalled a Met. expert once telling me that they had one such barometer on the top of the Air Ministry roof in London. Anyhow, I concluded, whichever one didn't like storms must be feeling a bit claustrophobic by now.

Our 'best-laid schemes' were to have the climbing on one day, preferably on the Saturday. This was to be covered by cameras at the summit camp, which looked directly across at the routes, and to have Jean-Frank Charlet ski down on the Sunday. This combination gave us several options. We could change things round or even, in the event of disastrous conditions, cram it all into one day, with due apologies to the viewers, of course.

Next morning was Transmission Day One and what a morning! 'The wind blew as 'twad blown its last.' Even so, by midday, the cameras rolled (or blinked) in the snowfall at the camera camp in the Allt a Mhuilinn. Eric Robson did his piece to camera and we were on the air, at least at a reduced altitude, and though the transmission from the summit didn't involve any climbing, the scenes were dramatic.

Anyone seeing the programme realized that it was a case of surviving, not entertaining. Murray Hamilton described the scene: "It was a bit unpleasant this morning. There was stuff coming across in the form of golfballs and doing about a thousand miles an hour." The weather was so bad that we just couldn't get the cameras into their scheduled positions and coupled up in time. It was depressing after all that effort.

That night a number of BBC engineers stayed on the summit in the tents. Mac kept them company, and for once it was mild and actually rained later. However, by 4 a.m., when all the sleeping-bags felt like damp poultices, frost clamped down. It didn't seem possible, but the weather grew worse. The press were hanging about Fort William willing something to happen.

Already the death toll on the mountains had raised a storm of criticism and our enterprise had been labelled foolhardy, despite the fact that we had rescued a number of people from the clutches of Ben Nevis over the past few weeks, and the Lazenby Inn had probably prevented further catastrophes.

On the Sunday, Saturday having been 'cancelled', I was concerned for people's safety, as we went up by treadmill once again to the summit. When I saw the camp, I was sure that exposure would hit someone in the not-too-distant future.

Some of the engineers already exhibited the tell-tale signs of exposure after an uncomfortable night and there was a zombie-like presence in the camp. Exposure is an insidious thing, and the signs are difficult to recognize early on. Later, disorientation and carelessness manifest themselves, along with faulty vision, and then that fatal lethargy. It can affect experienced climbers as well as those not used to such extreme conditions, but of course mountaineers know how to keep themselves warm and are not so likely to be early victims unless wet. I had spent 26 years teaching people winter climbing and could see the tell-tale signals.

Safety was my main responsibility, and I had a word with Mike on
the radio. "I don't like it. We may have to call it off." I said this, knowing
that several million people were geared up to watch our afternoon tele,
live from Ben Nevis.

"I was thinking the same thing from what I heard, Hamish. Shall we
do it now?"

"Let's give it an hour. I can't leave it later than that, because we've got
to get the engineers off the hill and it seems to be worsening."

"Right, Hamish, but I think we should pull out now. We wouldn't
have enough time to get set up an hour from now. It would be too close
to transmission."

"Right. That's it then."

I told the others at the summit camp. Some of the engineers were
keen to stay on and see the thing through, transmit and be damned. I
was more concerned in delivering eighteen cold engineers to base in
one piece than in leaving several corpses on the summit.

It was an unpopular decision to have to make, but with hindsight, it
was the correct one. Had anyone succumbed to exposure, we would
have been torn to shreds by the gaggle of press in Fort William and
rightly so.

Evacuating the BBC crew from the summit wasn't easy. They had to
get their crampons on and, as the bindings were frozen, this in itself
wasn't a piece of pie. One man who had lost his crampons was given
over to the sole charge of Dave Cuthbertson, who was one of the star
climbers on the Glencoe O.B., and as they set off into the grey-white
cloud, they looked a strange pair: the BBC man, large and falling on the
ice every other step, with Dave behind, slim and fragile-looking,
holding on to a short length of rope as though he were trying to control a
wild bull.

We put the other thirteen on a long rope. In front were two climbers
and at the rear three. Rab Carrington and John Yates were in the van
with compasses. Several of the BBC crew shed crampons on the way
down as the frozen bindings couldn't be tightened properly. However,
there was no stopping this human train once it had started, for
communication was virtually impossible in the blizzard, and it was a
case of getting the hell out of it as soon as possible. Half-way down the
mountain the helicopters with their landing lights on managed to come
in, bouncing in the potholes of the air. It was only 2 p.m.

"Sorry we didn't make it, Mike," I said to him when I got out of the
helicopter. "When it gets to that borderline state, you've got to make a

decision. You can't play with people's lives."

Mike's documentary proved to be very popular, so all wasn't lost. There was even some film of Charlet skiing and also about 50 minutes' live transmission from the top.

Dave Lazenby's squad came down the next day after sixteen days on the summit, but it wasn't until almost a week later that the storm puffed itself out, at least for long enough to get a Naval Sea King helicopter and P.L.M.'s two Jet Rangers to clear the camps.

In retrospect, it is heartening to know that the programme and the film caused so much public interest that the BBC were keen to attempt it the following year, but I for one was still licking my wounds from this last attempt and suggested to Phil a documentary on the Cocos Islands instead. "Some really good pirate stories — and treasure," I added. As it transpired, the Marines who had helped us had to go and fight the Falklands War, so we couldn't have tried again anyhow. An entry from the old Observatory Visitors' Book sums up our brush with the Ben:

> Roll by, thou dense and damp pea-soupy shroud,
> Do we thus reach the highest point in vain?
> Roll by! we say, and leave behind no cloud
> Our view to mar; but should'st thou still remain,
> Mark well the threat — Never shall we come again.